Pollinators: Life, Biology and Threats

Shyamasree Ghosh • Dhriti Banerjee
Editors

Pollinators: Life, Biology and Threats

Springer

Editors
Shyamasree Ghosh
School of Biological Sciences
National Institute of Science Education
and Research (NISER)
Jatni, Odisha, India

Dhriti Banerjee
Diptera
Zoological Survey of India (ZSI)
Kolkata, West Bengal, India

ISBN 978-981-95-0358-2 ISBN 978-981-95-0359-9 (eBook)
https://doi.org/10.1007/978-981-95-0359-9

This Springer imprint is published by the registered company Springer Nature Singapore Pte Ltd.
The registered company address is: 152 Beach Road, #21-01/04 Gateway East, Singapore 189721, Singapore

Preface

The global community of pollinators, including the immensely diverse insect Order of Diptera, plays a major role in food production by crops, plant reproduction, and other ecological services. The generalized pollinators include those reported from the alpine habitats including anthomyiids and tachinids and specialized pollinators including drosophilids, muscids, sarcophagids, and calliphorids. Syrphid and bombyliid flies like bees and butterflies sometimes visit similar flowers and are known to impact the plant fitness.

We discuss in this book, the biodiversity of pollinator Dipterans from the world and from India. We highlight the pollinator communities including Hover flies (syrphid Diptera), Indian thick headed flies, honey bee, bumblebees, solitary bees (Hymenoptera), and butterflies (Lepidoptera).

Despite importance of syrphid flies as pollinators, non-syrphid dipteran pollinators find importance in high altitudes and latitudes, including alpine and subarctic ecosystems with less abundant bee population. We discuss in brief the importance of non-syrphid flies as pollinators.

Due to their interactions with plants, they enable plant reproduction and growth. We highlight on the understanding on plant pollinator interactions.

In the recent times, several factors including global climate change, loss of habitat, altered use of land, and introduction of alien species has been a major threat to the life and survival of pollinator population leading to a substantial losses of pollinators which calls on an utmost need for conservation of wild pollinator communities. We discuss the major threats to the pollinator population and highlight impact of climate change on pollinator communities.

Honey bees have been studied for antiviral defense mechanism. We highlight insect immunity with a focus on the latest knowledge in the domain of antiviral defenses studied in honey bee.

Unique Features

1. We discuss the biodiversity of pollinator community from across the globe and India.
2. We highlight the importance of pollinator communities including Hover flies (syrphid Diptera), Indian thick headed flies, honey bee, bumblebees, solitary

bees (Hymenoptera), butterflies (Lepidoptera), and non-syrphid dipteran as pollinators.

3. We highlight the impact of climate change on life and survival of pollinator communities.
4. We discuss insect immunity highlighting antiviral defenses in honey bee.

Jatni, Odisha, India Shyamasree Ghosh
Kolkata, West Bengal, India Dhriti Banerjee

Contents

Editors and Contributors

About the Editors

Shyamasree Ghosh, MSc, PhD is Scientific Officer (F) at the School of Biological Sciences, National Institute of Science Education and Research (NISER), Bhubaneswar, Odisha, India. She is involved in teaching, research, and academic administration. She has previously worked as faculty at a postgraduate college affiliated with Bangalore University and held postdoctoral positions at Manipal Health Systems, Karnataka, and the Indian Association for the Cultivation of Science (IACS), Kolkata. She completed her BSc at Presidency College, University of Calcutta, and earned her MSc in Biotechnology and PhD in Science from the University of Calcutta and CSIR-IICB, respectively. Her research spans zoology, glycobiology, immunology, stem cells, computational biology, and nanotechnology. She has authored numerous peer-reviewed publications, books, chapters, and popular science articles. She serves as Associate Editor and editorial board member for reputed journals and is a reviewer for several international publications. A recipient of several academic awards, she has also contributed actively to science communication in multiple Indian languages.

Dhriti Banerjee, MSc, PhD is Scientist G and Director of the Zoological Survey of India (ZSI), Kolkata, West Bengal, India, under the Ministry of Environment, Forest and Climate Change. She is the first woman to head ZSI in its century-long history. She graduated from Presidency College, University of Calcutta, and earned her MSc in Zoology and PhD from the University of Calcutta and the University of Kalyani, respectively. With over two decades of experience at ZSI, her research interests include taxonomy, forensic dipterology, climate modeling, pollinator biology, and fly–plant interactions. She has published extensively in peer-reviewed journals and contributed to books and chapters. She has also led major digitization and data-basing initiatives at ZSI and has guided several PhD students.

Contributors

Dhriti Banerjee Diptera, Zoological Survey of India (ZSI), Kolkata, West Bengal, India

Succandra Chowdhury Post Graduate Department of Zoology, Bethune College, Kolkata, West Bengal, India

Shyamasree Ghosh School of Biological Sciences, National Institute of Science Education and Research (NISER), Jatni, Odisha, India

Md. Nurul Hasan Zoological Survey of India (Ministry of Environment, Forest and Climate Change, Government of India), Kolkata, West Bengal, India

Atanu Naskar Zoological Survey of India, New Alipore, Kolkata, West Bengal, India

Bristi Roy Zoological Survey of India, New Alipore, Kolkata, West Bengal, India

Jayita Sengupta Zoological Survey of India, New Alipore, Kolkata, West Bengal, India

Insect Pollinators: Biodiversity

1

Bristi Roy, Jayita Sengupta, Atanu Naskar,
Shyamasree Ghosh, and Dhriti Banerjee

Abstract

Pollination is a crucial ecosystem service, and insect pollinators are vital for maintaining global food security. This study investigates the contributions of various insect pollinators to pollination services, highlighting their importance in conserving this ecosystem service. Our research emphasizes the need for pollinator conservation strategies to ensure global food security.

Keywords

Pollinator · Insect · Biodiversity · Conservation · Food security

1.1 Introduction

Pollination is an essential part of sexual reproduction in flowering plants, where pollen grains from the anther of the male flower are transferred to the stigma of the female flower, enabling fertilization and the production of seeds. This crucial ecosystem service is provided by a wide variety of invertebrates and a few vertebrate taxa such as arthropods, birds, and small mammals. Plant-pollinator interactions are among the most important to the evolution of life on Earth (Bascompte and Jordano

B. Roy · J. Sengupta (✉) · A. Naskar
Zoological Survey of India, New Alipore, Kolkata, West Bengal, India

S. Ghosh
School of Biological Sciences, National Institute of Science Education and Research (NISER), Jatni, Odisha, India

D. Banerjee
Diptera, Zoological Survey of India (ZSI), Kolkata, West Bengal, India

S. Ghosh, D. Banerjee (eds.), *Pollinators: Life, Biology and Threats*,
https://doi.org/10.1007/978-981-95-0359-9_1

2007). Approximately 87.5% of flowering plants rely on animal pollinators, mostly insects (Ollerton 2017). Among various taxa that contribute uniquely to this essential ecological service, insects are the indispensable ones; they provide the greatest diversity, host specificity, and accessibility to even the tiniest flower. A plant and its insect pollinators share a mutualistic relationship where the plant provides food in the form of nectar and pollen, and the insect aids the plant's reproduction by facilitating pollination (Koneri et al. 2021).

Insect pollinators play an essential role in supporting both ecosystems and agricultural production, acting as agents of pollination for a wide variety of plants (Klein et al. 2007). These insects inadvertently transfer pollen between flowers of distant plants, which results in increased genetic diversity that may enhance the environmental tolerance of plants (Lundgren et al. 2015; Kumar et al. 2020). The economic value of insect pollination in agriculture is immense, as it not only increases crop yields but also improves the quality of produce (Cusser et al. 2016; Norfolk et al. 2016; Wietzke et al. 2018; Ahmad et al. 2021; Cohen et al. 2020). Approximately 35% of the global crop-dependent food production relies on pollinator guilds (Klein et al. 2007). By contributing to agricultural productivity, insect pollinators are key to maintaining food security and nutritional diversity worldwide. In addition to their agricultural value, insect pollinators are vital for biodiversity in various ways—many wild plant species, which form the foundation of ecosystems, depend on insect pollinators for reproduction and dispersal. Without these pollinators, plant populations would decline, which would, in turn, reduce the resources available for other species. By supporting the reproduction of these plant species, insect pollinators play an indirect yet significant role in creating and maintaining habitats for various animals, including birds and mammals. This interconnectedness makes pollinators vital in preserving biodiversity and sustaining healthy ecosystems.

Insect pollinators also provide essential ecosystem services that extend beyond pollination. Through their support of plant growth, they contribute to nutrient cycling and soil health, as decaying plant material enriches the soil with organic matter. Insect pollinators support plant reproduction and biodiversity, enhancing carbon sequestration, which helps mitigate climate change by absorbing carbon dioxide from the atmosphere.

Thus, the presence of healthy insect pollinator populations indirectly supports efforts to maintain balanced ecosystems and a stable climate. Beyond ecological and agricultural benefits, insect pollinators hold cultural and medicinal significance. Many medicinal plants rely on insect pollination, vital for natural and traditional medicines (Kozuharova et al. 2020; Samanta et al. 2023). Losing these pollinators would diminish access to valuable medicinal resources. Pollinators such as bees and butterflies also carry symbolic value, inspire art and culture, often representing resilience, community, and harmony in various cultures (Haynes 2013; Costa and Soares 2015; Duffus et al. 2021; Prendergast et al. 2021). Their presence enriches our natural surroundings, serving as reminders of the intricate connections within the environment and our dependence on it.

Some of the key insect pollinators include bees, butterflies, moths, beetles, wasps, and flies. Bees are considered the most effective pollinators due to their fidelity to specific flower types when foraging. Butterflies and moths also play essential roles, especially in pollinating plants with unique flower shapes or those that bloom at dawn or dusk. Beetles, wasps, and flies are vital for plants with distinctive scents, ensuring a diverse array of plants receive pollination. In the absence of these insects, many plants would face reproductive challenges, jeopardizing food production and ecosystem health.

1.2 Diversity and Contribution of Different Insect Taxa in Pollination

Contemporary pollinator diversity reflects millions of years of complex and varied coevolutionary interactions with both gymnosperms and angiosperms, involving both specialized and broad relationships that have shaped modern ecosystems (Ollerton 2017). Insect pollinators play a critical role in sustaining biodiversity and ensuring food security. While bees are the most celebrated pollinators, many other insects—including flies, beetles, butterflies, moths, and wasps—contribute to pollination in both natural and agricultural ecosystems. Among pollinating insects, the order Lepidoptera holds the highest diversity with more than 140,000 pollinating species, followed by Coleoptera with about 77,300 species, Hymenoptera with around 70,000 species, among true flies Diptera, hoverflies (Syrphidae) from two subfamilies approximately 5600 species, are well documented as pollinators (Katumo et al. 2022). Each taxonomic group has unique characteristics that influence its effectiveness in pollination, such as flower preferences, body structure, and foraging behaviors. The pollinator guild comprises the highest diversity. Understanding the contributions of different insect taxa can help develop strategies to protect and enhance pollinator diversity, which is crucial for ecosystem stability and agricultural productivity (Klein et al. 2007).

1.2.1 Lepidoptera (Butterflies & Moths)

The order Lepidoptera, comprising butterflies and moths, plays a significant yet often underrecognized role in pollination (Figs. 1.1, 1.2, 1.3, and 1.4). With around 160,000 known species, Lepidoptera exhibit a wide range of morphologies, behaviors, and floral preferences that make them important pollinators across various ecosystems (Powell 2009). Not all butterflies are flower-visiting; those that have mouthparts with long, thin, coiled proboscis adapted for feeding on a liquid diet are the flower visitors, e.g., all species of Danaids, Vanessas, Nymphalids, and males of Lycaenids, Papilios, Pierids, and Hesperiids (Reddi and Bai 1984). The tubular proboscis allows them to access nectar deep within flowers with long corollas, which many other pollinators cannot access. The lengths of proboscises vary greatly among species, with some, like the hawkmoth (*Manduca sexta*), possessing exceptionally long proboscises that enable them to pollinate flowers inaccessible to most insects (Raguso and Willis 2005). Lepidopteran offers both diurnal (butterflies) and nocturnal (moths) species, making them complementary pollinators for plants that have flowers open at specific times of day (Powell 2009). Moths, particularly sphinx

Fig. 1.1 Order Lepidoptera: *Junonia atlites* (Linnaeus 1763)—Grey Pansy. (Picture Courtesy: Debjani Ghosh and Director, ZSI)

Fig. 1.2 Order Lepidoptera: *Danaus chrysippus* (Linnaeus 1758)—Plain Tiger. (Picture Courtesy: Debjani Ghosh and Director, ZSI)

moths and hawkmoths, are strong fliers and are drawn to night-blooming flowers, which often have strong scents to attract them in low-light conditions.

Unlike bees, which often pollinate primarily for food gathering, Lepidoptera pollinate while seeking nectar, contributing to the reproductive success of numerous plant species.

Fig. 1.3 Order Lepidoptera: *Junonia almana* (Linnaeus 1758)—Peacock Pansy. (Picture Courtesy: Debjani Ghosh and Director, ZSI)

Fig. 1.4 Order Lepidoptera: *Phalanta phalantha* (Drury 1773)—Common Leopard. (Picture Courtesy: Debjani Ghosh and Director, ZSI)

1.2.1.1 General Characteristics

- They are medium to large-sized flying insects with bodies and wings that are completely covered by flat and overlapping scales and hairs.
- Antennae variable, often clavate or serrate, hooked or knobbed or plumose.
- Compound eyes are large with two or no ocelli.
- Siphoning-type mouthparts with a long, coiled proboscis.
- Wings have two pairs; often, the forewings are largest.

1.2.1.2 Floral Preference

As discussed earlier, due to the unique morphology of mouthparts (suctorial proboscis), moths and butterflies are fond of flowers with long corollas. Apart from this, Lepidopteran pollinators are much more host plant-specific than other pollinating insects. The floral preferences of butterflies are affected by color, structure, shape, and size of the flower, nectar quality, and quantity (Erhardt 1991; Boggs and Ross 1993; Ilse 1928; Yurtsever et al. 2010; Pohl et al. 2011; Tang et al. 2013; Tiedge and Lohaus 2017). Moreover, foraging preferences are influenced by the compatibility between flower morphology, such as corolla length, and butterfly morphology, like proboscis length (Tiple et al. 2009). For example, flowers with short corolla tubes are visited by Hesperidae butterflies, and Papilionidae butterflies visit the flowers with longer corolla tubes (Subedi et al. 2021). Yurtsever et al. (2010) show a high foraging preference for Lepidoptera in yellow and pink colors compared to red. The study conducted by Subedi et al. (2021) shows a significant preference of butterflies for herbaceous plants over woody plants, tubular flowers over nontubular flowers, and a higher butterfly abundance at yellow, white, and purple flowers than for pink flowers.

1.2.1.3 Major Pollinating Groups

Pollinating taxa of Lepidoptera are mainly in the moth families Sphingidae (hawk moths), Noctuidae (owlet moths), Geometridae (geometer moths), and the butterfly families Hesperiidae (skippers) and Papilionoidea (common butterflies). Common tiger (*Danaus genutia*), belonging to the family Nymphalidae (Danaidae), is one of the most common flower visitors among Lepidoptera, which often feed on milkweed (*Asclepias* spp.). Among moths, the Hawk moth (*Manduca sexta*), family Sphingidae, is known to visit and pollinate flowers such as the Sacred Datura (*Datura wrightii*), as its proboscis is well-adapted for reaching deep into tubular flowers (Alarcón et al. 2008).

1.2.1.4 Agricultural and Ecological Contribution

While butterflies are not as economically significant in crop pollination as bees, they do contribute to the pollination of certain fruit and seed crops. For instance, butterflies aid in the pollination of crops like cotton, blueberries, and certain herbs, especially the flowers pollinated by moths (phalaenophily) and butterflies (psychophily) are often deeply tubular to accommodate the long-coiled proboscis that is absent from most other insect pollinator groups (Fenster et al. 2004). Moths are essential nocturnal pollinators, particularly for plants that release their scents and bloom at night. Moths' attraction to sweetly scented flowers with pale or white colors makes them especially effective for plants adapted to night pollination. These plants include Asparagales (including Orchidaceae, Amaryllidaceae, Asparagaceae, and others), and the economically important *Elaeis guineensis* (Arecaceae), visited by large numbers of moths in the genus Pyroderces (Cosmopterigidae), and certain members of the Cactaceae family, which depend heavily on moths for successful pollination (MacGregor et al. 2015). They also play a role in the pollination of ornamental plants, which has value in horticulture and supports biodiversity in agricultural landscapes (Ollerton et al. 2011).

Apart from this, Lepidoptera contributes uniquely to sericulture, as no other order of insects can produce commercially viable silk fiber with the same structural integrity and desirable qualities as those from *Bombyx mori* (Goldsmith et al. 2005). The silk industry contributes significantly to the economy and GDP of many silk-producing countries. For example, India, one of the largest producers of silk, earned $213 million by exporting silk and silk products abroad between April 2023 and January 2024 (Reddy and Parasuramudu 2024).

Butterflies and moths also serve as indicator species, with butterflies being more suitable for analyzing open habitats and providing a more sensitive short-term indicator for conservation measures, while moths are effective indicators for both open and forested habitats (Rákosy and Schmitt 2011), meaning that their abundance and diversity reflect ecosystem health. Conservation efforts aimed at preserving Lepidoptera, such as creating pollinator habitats and reducing pesticide use, benefit these insects and support a wide range of other species that depend on Lepidoptera-pollinated plants. Conserving these pollinators is therefore essential for maintaining biodiversity and ecosystem resilience.

1.2.2 Hymenoptera (Bees and Wasps)

By the word "pollination," the very first thing we visualize in our mind is a bee nectaring on a flower, the reason behind this is their wide availability and diversity. The order Hymenoptera (Figs. 1.5, 1.6, 1.7, and 1.8) includes some of the most

Fig. 1.5 Order Hymenoptera: *Ceratina* sp. (Picture Courtesy: Debjani Ghosh and Bristi Roy and Director, ZSI)

Fig. 1.6 Order Hymenoptera: *Apis dorsata* (Fabricius 1793)—Rock Bee. (Picture Courtesy: Debjani Ghosh and Bristi Roy and Director, ZSI)

Fig. 1.7 Order Hymenoptera: *Xylocopa* sp.—Carpenter Bee. (Picture Courtesy: Debjani Ghosh and Bristi Roy and Director, ZSI)

effective and economically significant pollinators, such as bees, wasps, and ants, among these, bees are the most efficient insect pollinators due to their specialized structures for pollen collection and high fidelity to flowers, which maximize pollen transfer. Hymenoptera is the third most diverse insect order and the major contributor of pollination, especially social bees (*Apis* and *Bombus* spp.) are the dominant pollinators across every ecosystem (Wardhaugh 2015). Honey bees (*Apismellifera*)

Fig. 1.8 Order Hymenoptera: *Apis florea* (Fabricius 1787)—Red dwarf honey bee. (Picture Courtesy: Debjani Ghosh and Bristi Roy and Director, ZSI)

are globally managed, economically essential insect pollinators, providing pollination of natural, wild vegetation plants as well as agricultural crops (Calderone 2012; Garibaldi et al. 2013). Approximately 20,000 described bee species visit flowers and are potent pollinators (Naumann 1991). Unlike other pollinators that visit flowers primarily for nectar, bees actively collect pollen to feed their young, making them exceptionally effective in transferring pollen between flowers.

1.2.2.1 General Characteristics

- Minute to large-sized insects, social, parasitic even solitary.
- Antennae are 12-segmented in males and 13-segmented in females.
- Ocelli are generally present.
- Mouthparts adapted for biting, chewing lapping, and sucking.
- Wing two pairs, membranous and narrow.
- The abdomen is always with a basal pedicle.
- Female with conspicuous ovipositor modified into a saw, drill, or sting.

1.2.2.2 Floral Preference

The floral preferences of Hymenoptera play a crucial role in shaping plant-pollinator dynamics and influencing plant reproductive success. Bees, which represent the major pollinating taxa of Hymenoptera, show preferences for flowers with specific traits, such as bright colors (especially blue, purple, and yellow), shallow to moderately deep corollas, and abundant nectar and pollen. These traits align with the visual and olfactory cues that bees rely on for foraging (Giurfa et al. 1996). Bees are eusocial insects, perform a variety of movements termed as "dance" (e.g., *round*, *waggle*) to communicate the distance and direction of food sources to the other foraging members.

1.2.2.3 Major Pollinating Groups

(a) *Bees (Apidae, Megachilidae, Halictidae, Andrenidae)*: Bees are the primary pollinators among Hymenoptera. Honey bees (*Apis mellifera*) are critical agricultural pollinators globally, enhancing fruit set in crops like almonds, apples, and blueberries (Klein et al. 2007). Bumblebees (*Bombus* spp.) and solitary bees, such as leafcutter bees (*Megachile* spp.) and sweat bees (*Lasioglossum* spp.), also contribute to pollination across both wild plants and crops due to their foraging efficiency and behavior (Garibaldi et al. 2013). Solitary bees, in particular, are highly effective at pollinating specific plants due to their strong flower constancy.
(b) *Wasps (Vespidae, Scoliidae)*: Though less efficient than bees, certain wasps contribute to pollination, especially in tropical ecosystems. Some wasps, such as *Polistes* spp. (paper wasps), occasionally pollinate flowers while foraging for nectar (Brosi et al. 2008). Fig wasps (Agaonidae) are critical for the pollination of fig species (*Ficus* spp.), engaging in a mutualistic relationship where wasps pollinate figs while developing their larvae within the flowers (Cook and Rasplus 2003).
(c) *Ants (Formicidae)*: Ants are generally less significant as pollinators due to their limited mobility and tendency to remain on plants without transferring pollen over long distances. However, certain species are involved in pollination within specific habitats, such as dry or alpine ecosystems, where they can contribute to the pollination of low-growing plants by visiting flowers for nectar (Hölldobler and Wilson 1990).

1.2.2.4 Agricultural and Ecological Contribution

Honey bees are vital to agriculture as pollinators, contributing significantly to global food production. Their economic impact worldwide is valued between \$235 and \$285 billion annually (Lautenbach et al. 2012). Beyond pollination, bees hold significant ecological and economic roles that benefit both ecosystems and human society. One key contribution is their role in biodiversity support, as bees contribute to the maintenance of habitats by aiding plant growth, which in turn supports diverse animal communities that rely on these plants for food and shelter (Winfree 2010). Bees also have a part in natural pest control; honey bees and some solitary bees actively collect and remove resin and propolis (plant-derived substances) from their environments, which have antimicrobial properties, benefiting both bee colonies and plant health (Simone-Finstrom and Spivak 2010). Furthermore, bees produce valuable products for human use, such as honey, beeswax, and propolis. Honey is a major agricultural product with nutritional, medicinal, and economic value, supporting rural economies globally. Propolis and beeswax are also valued for their antibacterial and preservative properties, which find applications in medicine and cosmetics (García-Trejo et al. 2016). Additionally, bees serve as essential bioindicators for environmental health due to their sensitivity to changes in ecosystems, such as pesticide levels and climate change effects, providing early warning signs for ecological disturbances (Kevan 1999).

1.2.3 Coleoptera

Coleoptera, commonly known as beetles, comprise the largest order of insects, with approximately 400,000 species globally, representing about 25% of the world's identified species (Stork et al. 2015). Although commonly associated with herbivory or scavenging (Kirmse and Chaboo 2020), beetles are important pollinators, particularly in tropical and subtropical ecosystems (Ollerton 2017). Unlike bees or butterflies, which are renowned for their specialized pollination roles, beetles are generalized pollinators that often visit flowers for feeding, thus contributing to incidental pollination. Beetles are essential contributors to ecosystems, supporting the survival and diversity of numerous plant species (Ahrens et al. 2014). "Cantharophily" or beetle pollination is the oldest form of pollination, present since the early Cretaceous period (Bao et al. 2019), with fossil evidence of their interaction with ancient plant groups such as cycads and magnolias dating back to the Jurassic period (Thien 1980; Thien et al. 1996, 2000). This ancient association suggests that beetles played a crucial role in the evolution of flowering plants, influencing the development of floral characteristics adapted to beetle visitation.

1.2.3.1 General Characteristics

- Minute to large insects with leathery integuments.
- Antennae variable in shape and size, usually 11-segmented.
- Eyes conspicuous but ocelli absent.
- Mouthparts are biting and chewing type.
- Two pairs of wings, forewings or elytra, are similar to the body texture and are useless in flight.
- Cerciabsent.

1.2.3.2 Floral Preference and Major Pollinating Groups

Plants exclusively pollinated by beetles display four overlapping floral forms: painted bowl-shaped blossoms, brush-like structures, chambered flowers, and bilabiate (two-lipped) designs (Bernhardt 2000). Beetles contribute to pollination through unique methods compared to more specialized pollinators like bees, and also rely on odor to locate flowers in darkness (Gottsberger and Silberbauer-Gottsberger 1991).

Different plant families release distinct odors and floral scents tailored to attract specific beetle pollinators. For example, *Amorphophallus johnsonii* (Araceae) emits a carrion and dung-like odor to attract the beetle *Phaeochrous amplus*, a key pollinator of this species. Similarly, plants in the Annonaceae family, such as *Anaxagorea* sp., produce fruity smells that mimic the scent of ripening fruits, drawing in their beetle pollinators (Muinde and Katumo 2024). This phenomenon of plant-pollinator scent specialization is well-documented across various plant species and beetle groups. Some additional examples include:

(a) *Magnolia (Magnoliaceae)*: Magnolia species, such as *Magnolia grandiflora*, emit strong, sweet, and sometimes spicy odors to attract beetles, particularly *Megasoma elephas* (scarab beetle), which is known to be a primary pollinator for these plants. The strong, floral scent is essential for attracting beetles to their large, robust flowers (Pellmyr and Thien 1986).
(b) *Calycanthus (Calycanthaceae)*: *Calycanthus* species, such as *Calycanthus floridus*, release a sweet, fruity fragrance that appeals to beetles, which act as their primary pollinators. The scent is produced by the plant's floral tissues and helps attract beetles that feed on the pollen and contribute to the plant's reproductive process (Thien et al. 2000).
(c) *Cyclanthaceae*: In this family, plants like *Cyclanthus bipartitus* emit strong, fermented, or mold-like odors that attract specialized beetles, particularly *Platynus* species. These beetles are key pollinators for *Cyclanthus* flowers, which are typically bowl-shaped and designed to hold the beetles during pollination (Silberbauer-Gottsberger and Gottsberger 2003).
(d) *Annonaceae*: In addition to *Anaxagorea* sp., other plants in the Annonaceae family, like *Annona squamosa*, attract beetle pollinators with a blend of fruity and slightly decaying odors that mimic the scent of overripe fruits (Momose et al. 1998).

The major pollinating families of Coleoptera are Nitidulidae, Cerambycidae, Staphylinidae, Mordellidae, Curculionidae, Chrysomelidae, and Scarabaeidae (Muinde and Katumo 2024).

1.2.3.3 Agricultural and Ecological Contribution

Bawa (1990) ranked beetles as the fourth major insect pollinator, particularly important in tropical and subtropical habitats. Research has shown that various beetle species serve as key, and often exclusive, pollinators for several angiosperm plant families in subtropical and tropical regions, including Araceae, Arecaceae, Calycanthaceae, Cyclanthaceae, Annonaceae, Eupomatiaceae, and Magnoliaceae (Muinde and Katumo 2024). Although beetles are less commonly recognized in agricultural pollination, they contribute to the productivity of certain crops and medicinal plants. Beetle pollination is important for economically valuable plants such as palms, magnolias, and certain spice plants. Several commercially important crops and plants that rely on beetle pollination include cacao, palms (for oil and dates), and some spices. Their pollination services enhance yields, particularly in regions where beetle-pollinated crops are key agricultural products (Kearns et al. 1998). Beyond agriculture, beetle pollination provides ecosystem services that maintain floral diversity in wild landscapes. This, in turn, supports a wide array of wildlife, including

Fig. 1.9 Order Diptera: *Stomorhina* sp. (Picture Courtesy: Koustav Mukherjee and Director, ZSI)

herbivores and carnivores, by sustaining plant populations that are integral to food chains (Roubik 1995).

1.2.4 Diptera

Diptera or true flies constitute the fourth largest insect order (Figs. 1.9, 1.10, 1.11, and 1.12), providing a great array of morphological and biological diversity, and numerous groups that have evolved spectacular structural and behavioral adaptations commensurate with their environment (Kirk-Spriggs and Sinclair 2017). With approximately 155,000 reported species, Diptera represents 12% of all named insects on Earth (Grimaldi and Engel 2005). Diptera is the only order that has vertebrate parasites and a hematophagous species group, in addition to pollinators, decomposers, saprophagous, phytophagous, necrophagous, and even aquatic species groups. The family Syrphidae, or hoverflies, represents major Dipteran pollinators with about 5600 pollinating species in two subfamilies (Syrphinae & Eristalinae). Hoverflies are excellent mimics of bees and wasps, and display very bright and attractive coloration of the body, the tactics that enable them to circumvent predators.

1.2.4.1 General Characteristics

- Minute to medium-sized, having diverse adaptations—diurnal, nocturnal, terrestrial, and aquatic.
- Slender body with fragile integuments.
- Antennae variable, simple or aristate.

Fig. 1.10 Order Diptera: *Eristalis tenax* (Linnaeus 1758)—Hoverfly. (Picture Courtesy: Koustav Mukherjee and Director, ZSI)

Fig. 1.11 Order Diptera: *Melanostoma* sp.—Hoverfly. (Picture Courtesy: Koustav Mukherjee and Director, ZSI)

- Eyes are large and separate with three ocelli.
- Mouthparts show a range of adaptations—piercing, sucking, biting, lapping, and sponging.
- "True flies" with a single pair of wings, and hind wings reduced to halters act as a balancing organ during flight.

Fig. 1.12 Order Diptera: *Episyrphus balteatus* (De Geer 1776)—Hoverfly. (Picture Courtesy: Koustav Mukherjee and Director, ZSI)

1.2.4.2 Floral Preference

Hoverflies (Syrphidae) exhibit distinct floral preferences, often influenced by factors such as color, nectar availability, and floral morphology. Studies show that certain subfamilies within Syrphidae display specific color preferences. For example, Eristalinae exhibited a strong preference for white flowers, Pipizinae favored both white and yellow flowers, while Syrphinae showed less sensitivity to flower color (Klecka et al. 2018). Moreover, hoverfly species have also been reported to show innate floral color preference. For example, *Episyrphus balteatus* and *Eristalis tenax* have a preference for yellow flowers (Ilse 1949; Sutherland et al. 1999; Lunau 2014), *Rhingia campestris* and *Volucella bombylans* for blue flowers (Lunau 2014). Hoverflies also have a preference for open, accessible flowers, which allow easy access to nectar and pollen. They frequently visit flowers in families such as Apiaceae, Asteraceae, and Brassicaceae, which often feature shallow floral structures suited to the hoverfly's short mouthparts (Haslett 1989; Branquart and Hemptinne 2000). McFadden and Hans (2019) studied the foraging behavior of blow flies (*Calliphora terraenovae*) and showed a preference for purple flowers in both sexes. These floral preferences play a significant role in hoverfly ecology and their effectiveness as pollinators, contributing to pollination in both natural and agricultural ecosystems.

1.2.4.3 Major Pollinating Groups

Hoverflies are the major pollinating Diptera; apart from this, bee flies (Bombyliidae), Calliphoridae (blow flies), Tachinidae, and Bibionidae (March flies) also contribute significantly to the pollination of many agroecosystems and wild vegetations. We discuss this in greater detail in later chapters.

1.2.4.4 Agricultural and Ecological Contribution

Ssymank et al. (2008) have discussed the flower-visiting Dipterans to visit almost 555 flowering plants and more than 100 cultivated crops—mango, strawberry, cashew, tea, cacao, and onions, cauliflower, mustard, apples, carrots, leeks, and cassava; blow flies are commercially raised for pollination of above-mentioned plants as well as sunflower, buckwheat, canola, garlic, pepper, and lettuce and also greenhouse production; *Eristalis* sp. is used to pollinate peppers, *Musca* sp. and *Calliphora* sp. used to pollinate umbelliferous plants in greenhouse; *Eristalis* sp. and many other hoverflies are reared commercially for production of seeds for seed banks. Unlike bees, many dipteran pollinators are active under cooler and cloudier conditions, making them valuable pollinators in regions and seasons where bee activity is limited (Orford et al. 2015). Hoverflies provide dual ecosystem services—as pollinators in the adult stage and aphid predators during their larval stages, which provides natural pest control benefits that support integrated pest management strategies (Ssymank and Kearns 2009). Apart from this, many hoverfly species are migratory (*Episyrphus balteatus* and *Eupeodes* sp.), and can transport pollen over a long distance, thereby aiding in connecting geographically isolated populations through high levels of gene transfer (Doyle et al. 2020).

1.3 Discussion

Pollinator guild is in substantial decline over the past decades (van der Sluijs and Vaage 2016), and, among the declining pollinator groups, bees, dung beetles, and Lepidoptera are the taxa that have been affected the most (Lebuhn et al. 2013). The recent significant loss of pollinators due to habitat loss, disease outbreaks, pesticide exposure, altered usage of the land, and climate change (Potts et al. 2010; Goulson et al. 2015), led to a real need for land managers to conserve pollinator communities. Otherwise, the shortage of these pollinators will enhance the demand for agricultural land, a trend that will be more predicted in the developing world, potentially leading to decreased food security for the future world.

To address these declines, conservation strategies focus on habitat restoration, sustainable agricultural practices, and the reduction of pesticide use. Restoring native flowering habitats and implementing pollinator-friendly planting practices can enhance food and nesting resources. For instance, planting wildflower strips in agricultural landscapes can support both wild and managed pollinators by increasing floral diversity and habitat availability (Garibaldi et al. 2014). Reducing pesticide use or adopting pollinator-safe pesticides is another priority, as insecticides like neonicotinoids have been linked to declines in bee health and survival (Woodcock et al. 2017).

Climate change also poses a significant threat by altering flowering times and ranges, potentially disrupting plant-pollinator relationships. To mitigate these impacts, adaptive conservation measures, such as protecting climate-resilient habitats and fostering genetic diversity within pollinator populations, are essential (Kerr et al. 2015). Public awareness and policy support for pollinator conservation, including actions like the European Union's Pollinators Initiative, are equally important for fostering long-term conservation efforts.

References

Ahmad S, Khalofah A, Shahmshad Khan AS, Khan AK, Jilani MJ, Hussain T, Skalicky M, Ghramh HA, Ahmad Z (2021) Effects of native pollinator communities on the physiological and chemical parameters of loquat tree (*Eriobotrya japonica*) under open field condition. Saudi J Biol Sci 28:3235–3241. https://doi.org/10.1016/j.sjbs.2021.02.062

Ahrens D, Schwarzer J, Vogler AP (2014) The evolution of scarab beetles tracks the sequential rise of angiosperms and mammals. Proc R Soc B Biol Sci 281(1791):20141470

Alarcón R, Davidowitz G, Bronstein JL (2008) Nectar usage in a southern Arizona hawkmoth community. Ecology 89(6):1839–1847

Bao T, Wang B, Li J, Dilcher D (2019) Pollination of Cretaceous flowers. Proc Natl Acad Sci 116(49):24707–24711

Bascompte J, Jordano P (2007) Plant-animal mutualistic networks: the architecture of biodiversity. Annu Rev Ecol Evol Syst 38:567–593. https://doi.org/10.1146/annurev.ecolsys.38.091206.095818

Bawa KS (1990) Plant-pollinator interactions in tropical rain forests. *Annual review of Ecology and Systematics* 21: 399–422

Bernhardt P (2000) Convergent evolution and adaptive radiation of beetle-pollinated angiosperms. In: Pollen and pollination. Springer, pp 293–320

Boggs CL, Ross CL (1993) The effect of adult food limitation on life history traits in *Speyeria mormonia* (Lepidoptera: Nymphalidae). Ecology 74(2):433–441

Branquart E, Hemptinne JL (2000) Selective attraction of the aphidophagous hoverfly *Episyrphus balteatus* (Diptera: Syrphidae) to different plant species infested by aphids. Environ Entomol 29(3):622–627

Brosi BJ, Daily GC, Ehrlich PR (2008) Bees and pollination in Thailand. Biodivers Conserv 17(4):793–807

Calderone NW (2012) Insect pollinated crops, insect pollinators and US agriculture: trend analysis of aggregate data for the period 1992–2009. PLoS One 7(5):e37235

Cohen H, Philpott SM, Liere H, Lin BB, Jha S (2020) The relationship between pollinator community and pollination services is mediated by floral abundance in urban landscapes. Urban Ecosyst 24:275–290. https://doi.org/10.1007/s11252-020-01024-z

Cook JM, Rasplus JY (2003) Mutualists with attitude: coevolving fig wasps and figs. Trends Ecol Evol 18(5):241–248

Costa MF, Soares JC (2015) Free as a butterfly: symbology and palliative care. Rev Bras Geriatr Gerontol 18:631–641

Cusser S, Neff JL, Jha S (2016) Natural land cover drives pollinator abundance and richness, leading to reductions in pollen limitation in cotton agroecosystems. Agric Ecosyst Environ 226:33–42. https://doi.org/10.1016/j.agee.2016.04.020

Doyle T, Hawkes WLS, Massy R, Powney GD, Menz MHM, Wotton KR (2020) Pollination by hoverflies in the Anthropocene. Proc R Soc B Biol Sci 287(1927):20200508. https://doi.org/10.1098/rspb.2020.0508

Duffus NE, Christie CR, Morimoto J (2021) Insect cultural services: how insects have changed our lives and how can we do better for them. Insects 12(5):377

Erhardt A (1991) Flower preferences, nectar preferences and pollination effects of Lepidoptera. Adv Ecol 1:239–260

Fenster CB, Armbruster WS, Wilson P, Dudash MR, Thomson JD (2004) Pollination syndromes and floral specialization. Annu Rev Ecol Evol Syst 35:375–403. https://doi.org/10.1146/annurev.ecolsys.34.011802.132347

García-Trejo JF, Orantes-Bermejo JF, Antón-López J (2016) Uses of bee products in human health and nutrition. Nutrients 8(6):403

Garibaldi LA et al (2013) Wild pollinators enhance fruit set of crops regardless of honey bee abundance. Science 339(6127):1608–1611

Garibaldi LA, Carvalheiro LG, Vaissière BE, Gemmill-Herren B, Hipólito J, Freitas BM, Zhang H (2014) From research to action: enhancing crop yield through wild pollinators. Front Ecol Environ 12(8):439–447

Giurfa M, Núñez J, Chittka L, Menzel R (1996) Color preferences of flower-naïve honeybees. J Comp Physiol A 178(5):699–709

Goldsmith MR, Shimada T, Abe H (2005) The genetics and genomics of the silkworm, Bombyx mori. Ann Rev Entomol 50(1):71–100

Gottsberger G, Silberbauer-Gottsberger I (1991) Olfactory and visual attraction of Erioscelis emarginata (Cyclocephalini, Dynastinae) to the inflorescences of Philodendron selloum (Araceae). Biotropica 23:23–28

Goulson D, Nicholls E, Botías C, Rotheray EL (2015) Bee declines driven by combined stress from parasites, pesticides, and lack of flowers. Science 347(6229):1255957

Grimaldi D, Engel MS (2005) Evolution of the insects. Cambridge University Press

Haslett JR (1989) Interpreting patterns of resource utilization: randomness and selectivity in pollen feeding by adult hoverflies. Oecologia 78(4):433–442

Haynes D (2013) The symbolism and significance of the butterfly in ancient Egypt. Doctoral dissertation, Stellenbosch University, Stellenbosch

Hölldobler B, Wilson EO (1990) The ants. Harvard University Press

Ilse D (1928) Über den farbensinn der Tagfalter. Z Vgl Physiol 8(3–4):658–692

Ilse D (1949) Colour discrimination in the dronefly, *Eristalis tenax*. Nature 163:255–256

Katumo DM, Liang H, Ochola AC, Lv M, Wang QF, Yang CF (2022) Pollinator diversity benefits natural and agricultural ecosystems, environmental health, and human welfare. Plant Divers 44(5):429–435. https://doi.org/10.1016/j.pld.2022.01.005

Kearns CA, Inouye DW, Waser NM (1998) Endangered mutualisms: the conservation of plant-pollinator interactions. Annu Rev Ecol Syst 29:83–112

Kerr JT, Pindar A, Galpern P, Packer L, Potts SG, Roberts SM, Lazarus MC (2015) Climate change impacts on bumblebees converge across continents. Science 349(6244):177–180

Kevan PG (1999) Pollinators as bioindicators of the state of the environment: species, activity and diversity. Agric Ecosyst Environ 74(1–3):373–393

Kirk-Spriggs AH, Sinclair BJ (eds) (2017) Manual of Afrotropical Diptera. Volume 1. Introductory chapter and keys to Diptera families. Suricata 4. South African National Biodiversity Institute, Pretoria

Kirmse S, Chaboo CS (2020) Flowers are essential to maintain high beetle diversity (Coleoptera) in a Neotropical rainforest canopy. J Nat Hist 54(25–26):1661–1696

Klecka J, Hadrava J, Biella P, Akter A (2018) Flower visitation by hoverflies (Diptera: Syrphidae) in a temperate plant-pollinator network. *PeerJ* 6:e6025. https://doi.org/10.7717/peerj.6025. PMID: 30533311; PMCID: PMC6282941

Klein AM, Vaissière BE, Cane JH, Steffan-Dewenter I, Cunningham SA, Kremen C, Tscharntke T (2007) Importance of pollinators in changing landscapes for world crops. Proc R Soc B Biol Sci 274:303–313. https://doi.org/10.1098/rspb.2006.3721

Koneri R, Nangoy MJ, Wakhid (2021) Richness and diversity of insect pollinators in various habitats around Bogani Nani Wartabone National Park, North Sulawesi, Indonesia. Biodiversitas 22(1):288–297. https://doi.org/10.13057/biodiv/d220135

Kozuharova E, Aneva I, Goulson D (2020) Bees and medicinal plants – prospective for entomovectoring. In: Gao Y, Hokkanen H, Menzler-Hokkanen I (eds) Integrative biological control. Progress in biological control, vol 20. Springer, Cham. https://doi.org/10.1007/978-3-030-44838-7_15

Kumar P, Singh G, Singh H (2020) Impact of insect pollinators on quantitative and qualitative improvement in agricultural crops: a review. Int J Curr Microbiol App Sci 9(9):2359–2367
Lautenbach S, Seppelt R, Liebscher J, Dormann CF (2012) Spatial and temporal trends of global pollination benefit. PLoS One 7(4):e35954. https://doi.org/10.1371/journal.pone.0035954
Lebuhn G, Droege S, Connor EF, Gemmill-Herren B, Potts SG, Minckley RL, Griswold T, Jean R, Kula E, Roubik DW, Cane J, Wright KW, Frankie G, Parker F (2013) Detecting insect pollinator declines on regional and global scales. Conserv Biol 27(1):113–120
Lunau K (2014) Visual ecology of flies with particular reference to colour vision and colour preferences. J Comp Physiol A 200:497–512. https://doi.org/10.1007/s00359-014-0895-1
Lundgren R, Lazaro A, Totland Ø (2015) Effects of experimentally simulated pollinator decline on recruitment in two European herbs. J Ecol 103(2):328–337. https://doi.org/10.1111/1365-2745.12374
MacGregor CJ, Pocock MJ, Fox R, Evans DM (2015) Pollination by nocturnal Lepidoptera, and the effects of light pollution: a review. Ecol Entomol 40(3):187–198. https://doi.org/10.1111/een.12174
McFadden K, Hans K (2019) Foraging behavior of female blow flies (Diptera: Calliphoridae) based on visual cues. Instars J Stud Res 5:1–7
Momose K et al (1998) Pollination biology in a lowland dipterocarp forest in Sarawak, Malaysia. I. Characteristics of the plant-pollinator community. Am J Bot 85(10):1477–1501
Muinde J, Katumo DM (2024) Beyond bees and butterflies: the role of beetles in pollination system. J Nat Conserv 77:126523
Naumann ID (1991) Hymenoptera. In: The insects of Australia: a textbook for students and research workers, vol II. Melbourne University Press, Carlton, pp 916–1000
Norfolk O, Eichhorn MP, Gilbert F (2016) Flowering ground vegetation benefits wild pollinators and fruit set of almond within arid smallholder orchards. Insect Conserv Divers 9(3):236–243. https://doi.org/10.1111/icad.12162
Ollerton J (2017) Pollinator diversity: distribution, ecological function, and conservation. Annu Rev Ecol Evol Syst 48(1):353–376
Ollerton J, Winfree R, Tarrant S (2011) How many flowering plants are pollinated by animals? Oikos 120(3):321–326
Orford KA, Vaughan IP, Memmott J (2015) The forgotten flies: the importance of non-syrphid Diptera as pollinators. Proc R Soc B Biol Sci 282(1805):20142934
Pellmyr O, Thien LB (1986) Pollination ecology of the magnolias. Am J Bot 73(5):703–715
Pohl NB, Van Wyk J, Campbell DR (2011) Butterflies show flower colour preferences but not constancy in foraging at four plant species. Ecol Entomol 36(3):290–300. https://doi.org/10.1111/j.1365-2311.2011.01271.x
Potts SG, Biesmeijer JC, Kremen C, Neumann P, Schweiger O, Kunin WE (2010) Global pollinator declines: trends, impacts and drivers. Trends Ecol Evol 25(6):345–353
Powell JA (2009) Lepidoptera: moths, butterflies. In: Encyclopedia of insects. Academic Press, pp 559–587. https://doi.org/10.1016/B978-0-12-374144-8.00160-0
Prendergast KS, Garcia JE, Howard SR, Ren ZX, McFarlane SJ, Dyer AG (2021) Bee representations in human art and culture through the ages. Art Percept 10(1):1–62
Raguso RA, Willis MA (2005) Synergy between visual and olfactory cues in nectar feeding by naive hawkmoths, *Manduca sexta*. Anim Behav 69(2):407–418
Rákosy L, Schmitt T (2011) Are butterflies and moths suitable ecological indicator systems for restoration measures of semi-natural calcareous grasslandhabitats? Ecol Indic 11(5):1040–1045. https://doi.org/10.1016/j.ecolind.2010.10.010
Reddi CS, Bai GM (1984) Butterflies and pollination biology. Proc Anim Sci 93(4):391–396
Reddy YP, Parasuramudu M (2024) Role of sericulture in the Indian economy. J Res Appl Sci Biotechnol 3(4):107–116
Roubik DW (1995) Pollination of cultivated plants in the tropics. FAO Agricultural Services Bulletin

Samanta A, Bera B, Karmakar P (2023) Pollination ecology of an important medicinal plant *Hellenia speciosa* (J. Koenig) S.R. Dutta of Asiatic tropics. Biodiversitas 24(6):3152–3161. https://doi.org/10.13057/biodiv/d240608

Silberbauer-Gottsberger I, Gottsberger G (2003) Pollination ecology of tropical and temperate plants. Trends Plant Sci 8(5):190–195

Simone-Finstrom M, Spivak M (2010) Propolis and bee health: the natural history and significance of resin use by honey bees. Apidologie 41(3):295–311

Ssymank A, Kearns CA (2009) Flies—pollinators on two wings. Wings 32(2):3–7

Ssymank A, Kearns CA, Pape T, Thompson FC (2008) Pollinating flies (Diptera): a major contribution to plant diversity and agricultural production. Biodiversity 9(1–2):86–89

Stork NE, McBroom J, Gely C, Hamilton AJ (2015) New approaches narrow global species estimates for beetles, insects, and terrestrial arthropods. Proc Natl Acad Sci 112(24):7519–7523

Subedi B, Stewart AB, Neupane B, Ghimire S, Adhikari H (2021) Butterfly species diversity and their floral preferences in the Rupa Wetland of Nepal. Ecol Evol 11(5):2086–2099

Sutherland JP, Sullivan MS, Poppy GM (1999) The influence of floral character on the foraging behaviour of the hoverfly, Episyrphus balteatus. Entomol Exp Appl 93(2):157–164

Tang YC, Zhou CL, Chen XM, Zheng H (2013) Visual and olfactory responses of seven butterfly species during foraging. J Insect Behav 26:387–401

Thien LB (1980) Patterns of pollination in the primitive angiosperms. Biotropica 12:1–13

Thien LB et al (1996) The floral biology of the Magnoliaceae. In: Hunt D (ed) Magnolias and their allies. Proceedings of an international symposium. Royal Holloway, University of London, Egham, Surrey, pp 37–58

Thien LB, Azuma H, Kawano S (2000) New perspectives on the pollination biology of basal angiosperms. Int J Plant Sci 161(6):225–235. https://doi.org/10.1086/317575

Tiedge K, Lohaus G (2017) Nectar sugars and amino acids in day-and night-flowering Nicotiana species are more strongly shaped by pollinators' preferences than organic acids and inorganic ions. PLoS One 12(5):e0176865

Tiple AD, Khurad AM, Dennis RLH (2009) Adult butterfly feeding–nectar flower associations: constraints of taxonomic affiliation, butterfly, and nectar flower morphology. J Nat Hist 43(13–14):855–884

Van der Sluijs JP, Vaage NS (2016) Pollinators and global food Security: the Need for holistic global stewardship. Food Ethics 1:75–91. https://doi.org/10.1007/s41055-016-0003-z

Wardhaugh CW (2015) How many species of arthropods visit flowers? Arthropod Plant Interact 9(6):547–565. https://doi.org/10.1007/s11829-015-9398-4

Wietzke A, Westphal C, Gras P, Kraft M, Pfohl K, Karlovsky P, Pawelzik E, Tscharntke T, Smit I (2018) Insect pollination as a key factor for strawberry physiology and marketable fruit quality. Agric Ecosyst Environ 258:197–204. https://doi.org/10.1016/j.agee.2018.01.036

Winfree R (2010) The conservation and restoration of wild bees. Ann NY Acad Sci 1195(1):169–197

Woodcock BA, Bullock JM, Shore RF, Heard MS, Pereira MG, Redhead J, Pywell RF (2017) Country-specific effects of neonicotinoid pesticides on honey bees and wild bees. Science 356(6345):1393–1395

Yurtsever S, Okyar Z, Guler N (2010) What colour of flowers do Lepidoptera prefer for foraging? Biologia 65(6):1049–1056

Major Dipteran Pollinators: Hoverflies (Diptera: Syrphidae)

2

Bristi Roy, Jayita Sengupta, Atanu Naskar, Shyamasree Ghosh, and Dhriti Banerjee

Abstract

Syrphid flies, commonly referred to as hoverflies or flower flies, represent a taxonomically and ecologically diverse group within Diptera, playing vital roles as pollinators, biological control agents, and decomposers. Owing to their exceptional ecological plasticity and larval adaptability, hoverflies are distributed almost worldwide, with the exception of extreme deserts and Antarctica. This chapter synthesizes the existing knowledge on Indian syrphid diversity, highlighting that 368 species across 77 genera are known from the country, with Eristalinae being the most species-rich subfamily attributed to its wide range of larval feeding strategies. However, studies on the larval stages, especially those with significant ecological functions like decomposition and aphid predation, remain sparse in India. The importance of integrative research on Indian hoverfly fauna, incorporating taxonomy, molecular studies, and life history investigations, is strongly emphasized.

Keywords

Flower fly · Diversity · India · Distribution · Biological control

B. Roy · J. Sengupta (✉) · A. Naskar
Zoological Survey of India, New Alipore, Kolkata, West Bengal, India

S. Ghosh
School of Biological Sciences, National Institute of Science Education and Research (NISER), Jatni, Odisha, India

D. Banerjee
Diptera, Zoological Survey of India (ZSI), Kolkata, West Bengal, India

S. Ghosh, D. Banerjee (eds.), *Pollinators: Life, Biology and Threats*,
https://doi.org/10.1007/978-981-95-0359-9_2

2.1 Introduction

The family Syrphidae (Insecta: Diptera), commonly known as hoverflies or flower flies, is among the most diverse and widespread groups within the order Diptera. This remarkable diversity has enabled syrphids to colonize nearly every biotope globally, excluding Antarctica and extremely arid environments such as deserts (Ghahari et al. 2008). Syrphids are generally heliophilous, typically active during warm daylight hours. They are renowned for their exceptional mimicry of bees, wasps, and hornets both morphologically and behaviorally, a strategy that helps them evade predation. Like other Cyclorrhapha Diptera, syrphids undergo complete metamorphosis, comprising egg, three larval instars, pupa, and adult stages. Adults are readily identified by the presence of a characteristic spurious vein or "vena spuria" between R_{4+5} and M_{1+2} wing veins, a feature shared by most members of the family, with the notable exception of the genus *Graptomyza* Wiedemann, 1820 (Brunetti 1923).

Syrphids exhibit considerable variation not only in morphology but also in behavior, feeding habits, habitat preferences, and life history traits (Thompson and Rotheray 1998). Adult body sizes range from 4 to 25 mm, with coloration varying from black with yellow or orange markings on the thorax and abdomen, to more rarely observed hues such as brown, metallic green, or blue (Thompson and Rotheray 1998). Their body mass varies approximately 160-fold, from ~1 mg in *Neoascia* Williston, 1887, *Baccha* Fabricius 1805, and *Melanostoma* Schiner, 1860, to ~164 mg in females of *Volucella* Geoffroy, 1762 (Gilbert 1985). Moreover, some adult hoverflies exhibit innate floral color preferences. For instance, *Episyrphus balteatus* (De Geer, 1776) (Fig. 2.1) and *Eristalis tenax* (Linnaeus, 1758) prefer yellow flowers (Ilse 1949; Lunau 2014), while *Rhingia campestris* Meigen, 1822 and *Volucella bombylans* (Linnaeus, 1758) prefer blue (Lunau 2014). Klecka et al. (2018) demonstrated that floral color preferences differ among syrphid subfamilies: Eristalinae show the strongest preference for white flowers, Pipizinae prefer both white and yellow, while Syrphinae are less influenced by color cues.

Fig. 2.1 *Episyrphus (Episyrphus) balteatus* (De Geer, 1776), photo courtesy: Surajit Hazra

Historically, bees have been considered the most effective pollinators; however, global declines in bee populations linked to changes in land-use patterns, climate, and various anthropogenic factors have raised concerns about overdependence on bee-mediated pollination (Kluser and Peduzzi 2007; Van der Sluijs and Vaage 2016; Lima et al. 2022). Additional threats, including colony collapse disorder (Potts et al. 2010; UNEP 2010; Van der Sluijs et al. 2013; Wyss Institute 2016), have further impacted bee abundance. This scenario underscores the need for alternative pollinators. Non-bee pollinators such as moths, ants, beetles, and syrphid flies have gained recognition for their role in enhancing and stabilizing crop pollination services (Garibaldi et al. 2011). Wild pollinators can contribute independently or additively to crop yields alongside managed bee species (Garibaldi et al. 2013). Hoverflies, in particular, serve as effective pollinators across diverse habitats, often thriving in areas where bees and other Hymenoptera are scarce due to harsh environmental conditions (Warren et al. 1988). Unlike bees, hoverflies do not expend energy provisioning broods, which allows them to utilize less rewarding floral resources. They are active throughout the day (Brunetti 1923), and their pollination services may be more resilient to climate fluctuations than those of bees (Miličić et al. 2017). Additionally, there are migratory hoverfly species such as *E. balteatus* and *Eupeodes* spp. that contribute to long-distance pollen transport, promoting gene flow among geographically isolated plant populations (Doyle et al. 2020).

Larval feeding modes play a critical role in the ecological diversification of Syrphidae. Based on larval trophic strategies, Ghahari et al. (2008) categorized the family into three primary ecological groups: predators, miners, and decomposers (e.g., saprophagous larvae that thrive on decaying organic matter). These ecological groups correspond to various subfamilies: Syrphinae and Pipizinae larvae are entomophagous, preying mainly on soft-bodied arthropods such as aphids—an interaction extensively studied for its biocontrol potential. Microdontinae larvae are inquilines in ant nests, while Eristalinae include phytophagous (Rhingiini), mycophagous (Merodontini, Rhingiini), entomophagous (Volucellini), and saprophagous (Eristalini, Rhingiini, Volucellini) taxa (Thompson and Rotheray 1998). Such diversified larval feeding modes illustrate the ecological versatility of Syrphidae and their contribution to multiple ecosystem processes.

Beyond their extensively studied roles in pollination and aphidophagy from India (Ghorpadé 1981; Agarwala et al. 1984; Thompson and Ghorpadé 1992; Joshi and Ballal 2013; Singh et al. 2020; Khatua et al. 2020), hoverflies may also function as bio-decomposers, particularly through the saprophagous aquatic larvae of the tribe Eristalini, which inhabit organic pollutant-rich aquatic or semi-aquatic environments and facilitate waste decomposition and nutrient cycling (Campoy et al. 2017). Despite this ecological significance, studies on the biology and functional roles of eristaline syrphids (Figs 2.2, 2.3, and 2.4) are scarce, especially from the Oriental region, including India. To date, the early stages of only a few eristaline syrphids have been described from the Palaearctic, Neotropical, and Australian regions (Rotheray and Gilbert 1999; Pérez-Bañón et al. 2003a, b, 2013, 2020; Campoy et al. 2017, 2020). Consequently, parallel investigations from the Oriental region, particularly India, are lacking, leaving a substantial gap in understanding the biology and broader ecological roles of these taxa.

Fig. 2.2 *Eristalinus (Eristalinus) tabanoides* (Jaennicke 1867), photo courtesy: Jayita Sengupta

Fig. 2.3 *Eristalinus (Eristalinus) arvorum* (Fabricius 1787), photo courtesy: Jayita Sengupta

2.2 A Historical Perspective

Linnaeus, for the very first time in 1758 and 1761, described a total of 37 species of flower flies, based on an unspecified number of individuals per species (Thompson et al. 1982).

Afterward, research on flower flies was stretched forward by Fabricius (1775, 1781, 1805), with the introduction of several new genera and species. Much later, after the discovery of the flower fly species by Linnaeus, they were put together into a common taxon, "Family- Syrphidae" by Latreille in 1802. Before flower flies

Fig. 2.4 *Mesembrius (Mesembrius) bengalensis* (Wiedemann, 1819), photo courtesy: Surajit Hazra

were put together in a common taxon by Latreille, and afterward, several researchers around the world, like De Geer (1776), Wiedemann (1824), Macquart (1834, 1850, 1855), Walker (1849–1856), Schiner (1868), Kertész (1901), De Meijere (1908), Hervé-Bazin (1914, 1922, 1923), Sack (1928), Curran (1929), Greene (1949), Coe (1964), Hull (1941, 1942, 1944, 1950), Hippa (1974, 1978, 1985, 1990), Lambeck and van Brink (1973), Nielsen (2004), and Senior-White (1922), have enriched the Indian syrphid fauna. Among these, Brunetti's work is particularly noteworthy. In 1907, he published the first monograph on Indian Syrphidae, along with a comprehensive note on Oriental Diptera. Between 1908 and 1925, Brunetti extensively studied Indian syrphids, including the construction of taxonomic keys and systematic accounts, laying a foundational framework for future researchers.

Among Indian authors, the contributions of Biswas et al. (1975), Deoras (1943), Joseph & Parui (1972, 1973, 1976, 1977a, b, 1986), Joseph and Sharma (1976), Anand (1986), Kohli et al. (1988), Nayar (1968a, b), Singh et al. (1985), Ghorpade (1981, 1982, 1994, 2009, 2012, 2014), Mukherjee et al. (2006), Mitra et al. (2004, 2008a, b), and Shah et al. (2014) are highly regarded for significantly augmenting the knowledge of Indian Syrphidae, not only in terms of taxonomy but also concerning distributional records across various regions. More recently, Mitra et al. (2015), Sengupta et al. (2016a, 2017, 2018a, b, 2019a, b, c, d, 2020, 2023a, b, 2024), and Sankararaman et al. (2020, 2022) have made substantial contributions to this field.

In recent years, DNA barcoding has emerged as a powerful tool for hoverfly research in India. The application of molecular techniques, particularly DNA sequencing and barcoding, advances the understanding of Indian syrphid fauna by facilitating the description of novel species, associating different life stages, and ensuring accurate species identification within a traditional taxonomic framework (Anooj et al. 2020; Maqbool et al. 2024). Molecular phylogenetic analyses further complement these efforts by elucidating evolutionary relationships and refining the systematics of Indian Syrphidae (Kar et al. 2024). Karyotype and morphometric analysis of aphidophagous hoverflies have also been conducted in India in the recent past by Khajuria and Tripathy (2015).

2.3 Collection and Preservation of Hoverflies

Syrphid flies can be actively collected from the field during the daytime by using insect sweep nets. Additionally, passive trapping methods—particularly Malaise traps, pan traps, and UV pan traps—have proven highly effective in capturing large numbers of syrphid flies with minimal effort and time investment. Captured specimens need to be narcotized using ethyl acetate in killing jars. For transport and storage, specimens may be preserved either as dry collections in insect envelopes made of blotting paper or as wet collections in 70% ethanol. Prompt pinning of specimens is essential, as this preserves delicate morphological features that are often best observed when the specimen is fresh or recently immobilized. Specimens stored in insect envelopes require rehydration and relaxation before pinning. This is typically achieved by transferring the specimens into a relaxing chamber or desiccator. A relaxing chamber is a specialized glass container where a mixture of phenol and camphor (1:3 ratio) is placed at the base. Above this, a wire mesh and filter paper are positioned to allow the gradual absorption and diffusion of carbolic acid vapors. Specimens are placed above the filter paper and allowed to relax for at least 48 hours. This process facilitates softening of the body tissues and appendages, making them pliable for prompt pinning and mounting. After relaxation, specimens are pinned carefully, with wings, legs, and other taxonomically significant structures optimally stretched and positioned to allow clear visibility under a microscope during observation and morphological identification. Pinned specimens are subsequently transferred to a dry chamber for at least 72 hours, which helps eliminate residual moisture and ensures proper desiccation. These dry chambers should be equipped with chemical preservatives such as naphthalene to protect specimens from fungal growth, insect pests, and environmental factors like humidity and light exposure. Permanent computer-generated labels are affixed to each pinned specimen, providing critical collection data including state, district, locality, altitude, latitude, longitude, date of collection, and name of the collector. Once labeled, the specimens are placed into wooden entomological boxes or drawers lined with preservatives and stored in archival-quality cabinets for long-term preservation.

To ensure the longevity and integrity of the collection, regular maintenance of storage conditions is essential. This includes periodic replacement of chemical preservatives to prevent fungal and microbial contamination. Proper curation practices not only protect the physical integrity of the specimens but also ensure the scientific value of the collection for future taxonomic and ecological studies.

2.3.1 Habitat Range of Hoverflies

2.3.1.1 Forests and Woodlands

This type of habitat provides three to four layers of vegetation, namely, moss, herb, shrub, and tree layers. Additionally, forests and woodlands provide plenty of dead organic material in the form of dead wood and fallen leaves, serving as food for some hoverfly larvae. In different phases of the decay process, different hoverfly larvae feed on decaying wood. In deciduous forests, fallen and decaying leaves provide an additional source of decomposing material and are utilized by species of different genera, such as *Syritta* Le Peletier & Serville, 1828, and *Xylota* Meigen, 1822. The herb, shrub, and tree layer structure the forest vertically, providing hoverfly habitat at different heights. Hoverfly species belonging to genera *Platycherius* Le Peletier & Serville, 1828, *Episyrphus* Le Peletier & Serville, 1828, and *Melanostoma* Schiner, 1860, are usually found in the shrub and herb layers because their larvae mainly feed on aphids available in this layer. The varying light and shade conditions modify the forest habitats in many ways. In open areas and at the edge of the forest, more light reaches the forest floor. In these well-lit areas, lush herbaceous vegetation can grow, including flowering plants. The sun rays also heat these areas, providing relatively warm and dry spots. Such sites are preferred by hoverfly genera like *Epistrophe* Walker, 1852, and *Sphaerophoria* Le Peletier & Serville, 1828, as they bask in the sun and forage in flowers. In contrast, a dense canopy creates shady, damp conditions. These conditions are preferred by species from the genera like *Baccha* Fabricius, 1805 and *Sphegina* Meigen, 1822, as these species prefer the colder microclimate that such a site provides (Fig. 2.5).

2.3.1.2 Marshes, Peat Lands, and Bogs

This type of ecosystem provides important habitat for hoverflies. This type of ecosystem can contain areas of open water, water with emergent vegetation, peat-forming moss, damp grassland, and marsh forests. Some sites are rich in flowers, particularly those along banks and in damp grasslands. As succession occurs, the

Fig. 2.5 Forest

Fig. 2.6 Freshwater marsh

transition between the different wetland environments creates new habitat combinations. All of these provide ample habitat for both larvae and adult hoverflies. Particularly, all of the hoverfly species that have aquatic larval forms are found in this type of ecosystem. Species of genera like *Eristalis* Latreille, 1804, *Eristalinus* Rondani, 1845, and *Mesembrius* Rondani, 1857, were found to be abundant in such types of habitats, whereas some of them remain restricted to mesotrophic peat bogs. The decaying leaves and other vegetation occur in banks and peats along semi-aquatic conditions and form suitable habitats for larvae of genera like *Orthonevra* Macquart, 1829, whereas the adult hoverflies remain attracted to the vegetation and flowers on the bank (Fig. 2.6).

2.3.1.3 Meadows and Humid Grasslands

Many hoverflies are dependent on herbaceous vegetation and grasses, which is why meadows and humid grasslands are home to a lot of hoverfly species; therefore, such meadows and grasslands, when managed for agriculture, attract an ample diversity of syrphid fauna. This type of standing vegetation provides a microhabitat for both larval and adult hoverflies. Hoverfly genera like *Epysyrphus*, *Ischiodon* Sack, 1913, *Sphaerophoria, Platycherius,* and *Melanostoma* are typically found in grasslands as their larvae prey on aphids that live on herbs, sedges, and grasses. Thus, hoverfly fauna provides both pollination and pest control in this type of habitat. Hoverfly genera having comparatively large body sizes, e.g., *Eristalis, Eristalinus,* and *Mesembrius*, also prefer to visit agricultural landscapes as they might get more pollen or nectar reward from crops compared to wildflower patches (Fig. 2.7).

Fig. 2.7 Agricultural land

2.3.1.4 Xerothermic Grasslands and Rocks

This type of ecosystem includes rocky grounds as well as dry habitat areas where the intensity of temperature is high due to the sun's radiation. These types of habitats contain species specially adapted to xerothermic environments. Such xeric inhabitant hoverfly species include genera like *Eumerus* Meigen, 1822, *Merodon* Meigen, 1803, and *Volucella* Geoffroy, 1762, among these, *Volucella* survives in rocky habitats as its larvae live in the nests of wasps (Fig. 2.8).

2.3.1.5 Gardens and Residential Areas

Residential areas contain many green spaces; these spaces have characteristics of different natural environments. Here, areas of trees and shrubs are interspersed with grasslands and warm rocky areas. Gardens usually contain many hoverfly species, including *Episyrphus*, *Melanostoma, Platycherius, Sphaerophoria, Eristalis, Eristalinus, Mesembrius, Volucella,* and *Cheilosia* Meigen, 1822. On the other hand, parks and larger green areas contain a wide range of hoverfly species. Such sites have the highest species richness of all residential areas (Fig. 2.9).

2.4 Biology

Syrphid flies are sexually dimorphic, with males having contiguous or holoptic or closely approximated eyes and females having dichoptic or separated eyes. Hoverflies are holometabolous; their life cycle includes four stages: egg, larva, pupa, and adult. Their eggs look like tiny rice grains, laid both singly and in clutches

Fig. 2.8 Rocky streamside vegetation zone

Fig. 2.9 Garden

on leaves, decaying plant tissue, adjacent areas of polluted water bodies, drainages, etc. After hatching, the larval stage passes through three instars. The larvae are amphipneustic, slug-like, with reduced head capsules. Larvae that are adapted to live aquatic lives have an elongated posterior tail-like breathing siphon that helps them to breathe air while submerged. Syrphid pupae are adecticous and coarctated types, which means pupae are without articulated mandibles and are enclosed in the hardened cuticle of the penultimate larval instar called a puparium.

2.4.1 Differential Feeding Modes of Syrphid Larvae and Associated Ecological Functions

2.4.1.1 Aphidophagy

Larvae of the subfamily Syrphinae are predatory on soft-bodied insects, mainly aphids, thereby aiding in controlling the aphid population in the agricultural landscape. Major aphidophagous genera to provide potential biocontrol are *Allograpta* Osten Sacken 1875a, *Scaeva* Fabricius, 1805, *Syrphus* Fabricius, 1775, *Eupeodes* Osten Sacken, 1877, and *Sphaerophoria*.

2.4.1.2 Phytophagy

Phytophagous hoverflies, particularly those belonging to the genera *Cheilosia*, *Eumerus*, and *Merodon*, exhibit larval stages that feed on plant tissues such as roots, bulbs, stems, or leaves. Their close association with specific host plants and habitats makes them highly sensitive to shifts in vegetation composition, land-use practices, and habitat fragmentation. Consequently, these genera can serve as effective bioindicators for monitoring landscape changes, providing insights into ecosystem health, agricultural intensification, and the long-term impacts of anthropogenic disturbance on plant–insect interactions.

2.4.1.3 Myrmecophagy

Members of the subfamily Microdontinae are inquilines, lay eggs on ant nests, and remain close to ant nests. After hatching, their larvae feed on dead and immature ants, thereby, serving as a potential scavenger at the ant nests.

2.4.1.4 Saprophagy

Larvae of the subfamily Erstalinae (e.g., *Eristalis*, *Eristalinus*, *Palpada* Macquart, 1834, *Helophilus* Meigen, 1822) thrive on organic pollutant-loaded aquatic or semiaquatic media such as dirty ponds, bogs, manure lagoons, municipal sewage discharge, and drainage, where they consume decaying organic matter and thereby aid in rapid decomposition and nutrient cycling in aquatic media. Their potential to reduce organic wastes can be utilized in wastewater management.

2.5 Systematic List of Indian Syrphid Fauna

Family SYRPHIDAE
Subfamily MICRODONTINAE
Tribe MICRODONTINI
Genus *Indascia* Keiser, 1958
  Indascia brachystoma (Wiedemann, 1824)
Genus *Metadon* Reemer and Ståhls, 2013
  Metadon annandalei (Brunetti, 1907)
  Metadon auricinctus (Brunetti, 1908)
  Metadon flavipes (Brunetti, 1908)

Metadon ruficaudus (Brunetti, 1907)
Metadon squamipennis (Brunetti, 1923)

Genus *Microdon* Meigen, 1803

Microdon carbonarius Brunetti, 1923
Microdon unicolor Brunetti, 1915

Subgenus *Chymophila* Macquart, 1834

Microdon (Chymophila) stilboides Walker, 1849

Subgenus *Microdon* Meigen, 1803

Microdon (Microdon) apidiformis Brunetti, 1924
Microdon (Microdon) bellus Brunetti, 1923
Microdon (Microdon) metallicus De Meijere, 1914

Genus *Paramixogaster* Brunetti, 1923

Paramixogaster brunettii Reemer and Ståhls, 2013
Paramixogaster contractus (Brunetti, 1923)
Paramixogaster conveniens (Brunetti, 1923)

Tribe SPHEGINOBACCHINI

Genus *Spheginobaccha* Meijere, 1908

Spheginobaccha chillcotti Thompson, 1974
Spheginobaccha macropoda (Bigot, 1884b)

Subfamily ERISTALINAE

Tribe BRACHYOPINI

Genus *Lejogaster* Rondani, 1857

Lejogaster tarsata (Meigen, 1822)

Genus *Myolepta* Newman, 1838

Subgenus *Myolepta* Newman, 1838

Myolepta (Myolepta) himalayana Brunetti, 1915

Genus *Orthonevra* Macquart, 1829

Orthonevra aenethorax Kohli, Kapoor & Gupta, 1988
Orthonevra indica (Brunetti, 1915)
Orthonevra kozlovi Stackelberg, 1952
Orthonevra nobilis (Fallén, 1817)

Genus *Sphegina* Meigen, 1822

Subgenus *Asiosphegina* Stackelberg, 1974

Sphegina (Asiosphegina) bispinosa Brunetti, 1915
Sphegina (Asiosphegina) carinata Hippa, Steenis & Mutin, 2015
Sphegina (Asiosphegina) ghatsi Steenis, Hippa & Mutin, 2018
Sphegina (Asiosphegina) hamulata Steenis, Hippa & Mutin, 2018
Sphegina (Asiosphegina) javana Meijere, 1914
Sphegina (Asiosphegina) karnataka Steenis, Hippa & Mutin, 2018
Sphegina (Asiosphegina) radula Hippa, Steenis & Mutin, 2015
Sphegina (Asiosphegina) setosa Steenis, Hippa & Mutin, 2018
Sphegina (Asiosphegina) tristriata Brunetti, 1913

Subgenus *Sphegina* Meigen, 1822

Sphegina (Sphegina) asciiformis Brunetti, 1915
Sphegina (Sphegina) guptai Mutin, 1998

Sphegina (Sphegina) kumaoniensis Mutin, 1998
Sphegina (Sphegina) tricoloripes Brunetti, 1915

Tribe CALLICERINI

Genus *Callicera* Panzer, 1806

Callicera christiani Ghorpade, 1982
Callicera nitens Coe, 1964
Callicera robusta Coe, 1964

Tribe CERIOIDINI

Genus *Ceriana* Fabricius, 1794

Ceriana brevis (Brunetti, 1923)
Ceriana compacta (Brunetti, 1907)
Ceriana dimidiatipennis (Brunetti, 1923)
Ceriana ornatifrons (Brunetti, 1915)

Genus *Monoceromyia* Shannon, 1922

Monoceromyia crux (Brunetti, 1915)
Monoceromyia eumenioides (Saunders, 1842)
Monoceromyia fenestrata (Brunetti, 1923)
Monoceromyia flavipennis (Meijere, 1908)
Monoceromyia flavoscutata Sankararaman, Anooj & Mengual, 2022
Monoceromyia himalayensis (Meijere, 1908)
Monoceromyia javana (Wiedemann, 1824)
Monoceromyia nigra Sankararaman, Anooj & Mengual, 2022
Monoceromyia obscura (Brunetti, 1907)
Monoceromyia patricia (Brunetti, 1923)
Monoceromyia polistoides (Brunetti, 1923)
Monoceromyia tredecimpunctata (Brunetti, 1923)
Monoceromyia trinotata (De Meijere, 1914)

Genus *Sphiximorpha* Rondani, 1850

Sphiximorpha fruhstorferi (Meijere, 1908)
Sphiximorpha fulvescens (Brunetti, 1915)
Sphiximorpha triangulifera (Brunetti, 1913)

Tribe ERISTALINI

Genus *Eristalinus* Rondani, 1845

Subgenus *Eristalinus* Rondani, 1845

Eristalinus (Eristalinus) aeneus (Scopoli, 1763)
Eristalinus (Eristalinus) arvorum (Fabricius, 1787)
Eristalinus (Eristalinus) aurulans (Wiedemann, 1824)
Eristalinus (Eristalinus) megacephalus (Rossi, 1794)
Eristalinus (Eristalinus) obliquus (Wiedemann, 1824)
Eristalinus (Eristalinus) polychromata (Brunetti, 1923)
Eristalinus (Eristalinus) quadristriatus (Macquart, 1846)
Eristalinus (Eristalinus) quinquestriatus (Fabricius, 1794)
Eristalinus (Eristalinus) sepulchralis (Linnaeus, 1758)
Eristalinus (Eristalinus) tabanoides (Jaennicke, 1867)
Eristalinus (Eristalinus) tarsalis (Macquart, 1855)

Subgenus *Eristalodes* Mik, 1897

Eristalinus (Eristalodes) paria (Bigot, 1880)

Eristalinus (Eristalodes) taeniops (Wiedemann, 1818)

Subgenus *Merodonoides* Curran, 1931

Eristalinus (Merodonoides) fasciatus (Macquart, 1834)

Genus *Eristalis* Latreille, 1804

Subgenus *Eoseristalis* Kanervo, 1938

Eristalis (Eoseristalis) albibasis Bigot, 1880

Eristalis (Eoseristalis) angustimarginalis Brunetti, 1923

Eristalis (Eoseristalis) arbustorum (Linnaeus, 1758)

Eristalis (Eoseristalis) brevifacies Coe, 1964

Eristalis (Eoseristalis) cerealis Fabricius, 1805

Eristalis (Eoseristalis) curvipes Schiner, 1868

Eristalis (Eoseristalis) himalayensis Brunetti, 1908

Eristalis (Eoseristalis) intricarioides Brunetti, 1923

Eristalis (Eoseristalis) suturalis Brunetti, 1923

Subgenus *Eristalis* Latreille, 1804

Eristalis (Eristalis) tenax (Linnaeus, 1758)

Genus *Helophilus* Meigen, 1822

Subgenus *Helophilus* Meigen, 1822

Helophilus (Helophilus) trivittatus (Fabricius, 1805)

Genus *Kertesziomyia* Shiraki, 1930

Subgenus *Kertesziomyia* Shiraki, 1930

Kertesziomyia (Kertesziomyia) cyanea (Brunetti, 1913)

Subgenus *Pseuderistalis* Shiraki, 1930

Kertesziomyia (Pseuderistalis) fascipennis Thompson, 1975

Kertesziomyia (Pseuderistalis) nigra (Wiedemann, 1824)

Genus *Mallota* Meigen, 1822

Mallota orientalis (Wiedemann, 1824)

Mallota varicolor (Walker, 1856)

Mallota vilis (Wiedemann, 1830)

Genus *Mesembrius* Rondani, 1857

Subgenus *Mesembrius* Rondani, 1857

Mesembrius (Mesembrius) bengalensis (Wiedemann, 1819)

Mesembrius (Mesembrius) quadrivittatus (Wiedemann, 1819)

Mesembrius (Mesembrius) sharpi (Kohli, Kapoor & Gupta, 1988)

Mesembrius (Mesembrius) tuberculatus (Brunetti, 1907)

Mesembrius (Mesembrius) vestitus (Wiedemann, 1824)

Genus *Pararctophila* Hervé-Bazin, 1914

Pararctophila bengalensis Kohli, Kapoor & Gupta, 1988

Pararctophila oberthuri Hervé-Bazin, 1914

Genus *Phytomia* Guérin, 1833

Subgenus *Phytomia* Guérin, 1833

Phytomia (Phytomia) aesymnus (Walker, 1849)

Phytomia (Phytomia) argyrocephala (Macquart, 1842)

Phytomia (Phytomia) chrysopyga (Wiedemann, 1819)
Phytomia (Phytomia) errans (Fabricius, 1787)
Phytomia (Phytomia) zonata (Fabricius, 1787)

Subgenus *Dolichomerus* Macquart, 1850
Phytomia (Dolichomerus) crassa (Fabricius, 1787)

Genus *Pseudovolucella* Shiraki, 1930
Pseudovolucella eristaloides (Brunetti, 1913)
Pseudovolucella himalayensis (Brunetti, 1907)
Pseudovolucella hingstoni Coe, 1964

Genus *Tigridemyia* Bigot, 1882b
Tigridemyia curvigaster (Macquart, 1842)

Tribe MERODONTINI

Genus *Azpeytia* Walker, 1865
Azpeytia bifascia Brunetti, 1907

Genus *Eumerus* Meigen, 1822
Eumerus aeneithorax Brunetti, 1915
Eumerus albifrons Walker, 1852
Eumerus ammophilus Paramonov, 1927
Eumerus argentipes Walker, 1861
Eumerus aurifrons (Wiedemann, 1824)
Eumerus figurans Walker, 1859
Eumerus halictoides Brunetti, 1915
Eumerus kashmerensis Kohli, Kapoor & Gupta, 1988
Eumerus nepalensis Brunetti, 1908
Eumerus nicobarensis Schiner, 1868
Eumerus perpensus Brunetti, 1917
Eumerus perplexus Brunetti, 1917
Eumerus pulcherrimus Brunetti, 1915
Eumerus pulverulentus Brunetti, 1923
Eumerus rufoscutellatus Brunetti, 1913
Eumerus sexvittatus Brunetti, 1915
Eumerus vestitus Bezzi, 1912

Tribe MILESIINI

Genus *Blera* Billberg, 1820
Blera himalaya Thompson, 2000

Genus *Chalcosyrphus* Curran, 1925

Subgenus *Cheiroxylota* Hull, 1949
Chalcosyrphus (Cheiroxylota) dimidiatus (Brunetti, 1923)

Subgenus *Syrittoxylota* Hippa, 1978
Chalcosyrphus (Syrittoxylota) annulatus (Brunetti, 1913)

Subgenus *Xylotina* Hippa, 1978
Chalcosyrphus (Xylotina) ornata (Brunetti, 1915)

Genus *Criorhina* Meigen, 1822
Criorhina bomboides Hull, 1944
Criorhina crioarctos Hull, 1944

Criorhina excavata Curran, 1929
Criorhina imitator Brunetti, 1915
Criorhina pallidipes Curran, 1929
Criorhina pallipilosa Hull, 1944
Criorhina rubropilosa Hull, 1950
Criorhina simioides (Brunetti, 1908)
Criorhina spinitarsis Curran, 1929
Criorhina tripilosa Coe, 1964
Criorhina vivida Brunetti, 1923

Genus *Lycastris* Walker, 1856
Lycastris albipes Walker, 1856
Lycastris austeni Brunetti, 1923
Lycastris flavohirta Brunetti, 1907
Lycastris griseipennis Coe, 1964

Genus *Matsumyia* Shiraki, 1949
Matsumyia dentata (Brunetti, 1908)

Genus *Milesia* Latreille, 1804
Milesia balteata Kertész, 1901
Milesia brunneonigra Hippa, 1990
Milesia caesarea Hippa, 1990
Milesia cinnamomea Hippa, 1990
Milesia cretosa Hippa, 1990
Milesia ferruginosa Brunetti, 1913
Milesia illustris Hippa, 1990
Milesia mima Hippa, 1990
Milesia querini Ghorpadé, 2014
Milesia semifulva De Meijere, 1914
Milesia sexmaculata Brunetti, 1915
Milesia variegata Brunetti, 1908
Milesia verticalis Brunetti, 1923

Genus *Spilomyia* Meigen, 1803
Spilomyia manicata (Rondani, 1865)

Genus *Syritta* Le Peletier & Serville, 1828
Syritta fasciata (Wiedemann, 1830)
Syritta indica (Wiedemann, 1824)
Syritta orientalis Macquart, 1842
Syritta pipiens (Linnaeus, 1758)
Syritta proximata Lyneborg & Barkemeyer, 2005
Syritta stylata Lyneborg & Barkemeyer, 2005
Syritta triangulifera Keiser, 1958

Genus *Temnostoma* Le Peletier & Serville, 1828
Temnostoma nigrimanus Brunetti, 1915

Genus *Xylota* Meigen, 1822
Xylota auronitens Brunetti, 1908
Xylota bistriata Brunetti, 1915

Xylota carbonaria Brunetti, 1923
Xylota cuprina Bigot, 1885
Xylota nursei Brunetti, 1923
Xylota penicillata Brunetti, 1923

Genus *Brachypalpoides* Hippa, 1978
Brachypalpoides cupreiventris (Brunetti, 1923)
Brachypalpoides discolor Hippa, 1985
Brachypalpoides dives (Brunetti, 1908)
Brachypalpoides makiana (Shiraki, 1930)
Brachypalpoides plumipes Hippa, 1985

Genus *Cheilosia* Meigen, 1822
Cheilosia apicalis Brunetti, 1913
Cheilosia corydon (Harris, 1780)
Cheilosia himalayensis (Brunetti, 1915)
Cheilosia hirticincta Brunetti, 1915
Cheilosia kalatopensis Nayar, 1968a, b
Cheilosia nigroaenea Brunetti, 1915
Cheilosia pilipes (Bigot, 1884b)
Cheilosia plumbiventris (Brunetti, 1915)

Subgenus *Montanocheila* Barkalov, 2002
Cheilosia (Montanocheila) christiani Mengual & Barkalov, 2023
Cheilosia (Montanocheila) songarea Becker, 1894

Genus *Korinchia* Edwards, 1915
Korinchia rufa Hervé-Bazin, 1922

Tribe RHINGIINI

Genus *Endoiasimyia* Bigot, 1882b
Endoiasimyia indiana Bigot, 1882b

Genus *Ferdinandea* Rondani, 1844
Ferdinandea isabella Hull, 1942
Ferdinandea longifacies Coe, 1964
Ferdinandea montana Hull, 1942

Genus *Rhingia* Scopoli, 1763

Subgenus *Rhingia* Scopoli, 1763
Rhingia (Rhingia) angusticincta Brunetti, 1908
Rhingia (Rhingia) binotata Brunetti, 1908
Rhingia (Rhingia) cincta De Meijere, 1914
Rhingia (Rhingia) laticincta Brunetti, 1907
Rhingia (Rhingia) semicinerea Brunetti, 1923
Rhingia (Rhingia) sexmaculata Brunetti, 1913
Rhingia (Rhingia) siwalikensis Nayar, 1968b

Tribe VOLUCELLINI

Genus *Graptomyza* Wiedemann, 1820
Graptomyza angustimarginata Brunetti, 1923
Graptomyza brevirostris Wiedemann, 1820
Graptomyza flavonotata Brunetti, 1917

Graptomyza hardyi Greene, 1949
Graptomyza longirostris Wiedemann, 1820
Graptomyza nigripes Brunetti, 1913
Graptomyza sexnotata Brunetti, 1908
Graptomyza tinctovittata Brunetti, 1915
Graptomyza ventralis Wiedemann, 1830

Genus *Volucella* Geoffroy, 1762
Volucella basalis Brunetti, 1907
Volucella decorata Walker, 1859
Volucella discolor Brunetti, 1908
Volucella flavoscutella Sack, 1928
Volucella kinnaurensis Sengupta, Naskar & Banerjee, 2023b
Volucella lividiventris Brunetti, 1908
Volucella peleterii Macquart, 1834
Volucella pellucens (Linnaeus, 1758)
Volucella ruficauda Brunetti, 1907
Volucella ursina De Meijere, 1914
Volucella varipila Coe, 1964

Subfamily PIPIZINAE

Genus *Heringia* Rondani, 1856
Heringia cyanea (Brunetti, 1915)

Subfamily SYRPHINAE

Tribe MELANOSTOMINI

Genus *Melanostoma* Schiner, 1860
Melanostoma orientale (Wiedemann, 1824)
Melanostoma pedium (Walker, 1852)
Melanostoma univittatum (Wiedemann, 1824)

Genus *Xanthandrus* Verrall, 1901
Xanthandrus garhwalensis (Kohli, Kapoor & Gupta, 1988)

Subgenus *Xanthandrus* Verrall, 1901
Xanthandrus (Xanthandrus) indicus Curran, 1933

Tribe BACCHINI

Genus *Baccha* Fabricius 1805
Baccha maculata Walker, 1852
Baccha plumbicincta Brunetti, 1915

Genus *Platycheirus* Le Peletier & Serville, 1828
Platycheirus albimanus (Fabricius, 1781)
Platycheirus ambiguum (Fallén, 1817)
Platycheirus angustatus (Zetterstedt, 1843)
Platycheirus cryophilus Nielsen, 2007
Platycheirus himalayensis Brunetti, 1915
Platycheirus kashmiricus Nielsen, 2004
Platycheirus peteri (Doesburg, 1955)

Subgenus *Tuberculanostoma* Fluke, 1943
Platycheirus (Tuberculanostoma) solitarius (Doesburg, 1955)

Genus *Rohdendorfia* Smirnov, 1924
Rohdendorfia bella Mengual & Barkalov, 2019
Tribe SYRPHINI
Genus *Allobaccha* Curran, 1928
Allobaccha binghami Ghorpadé, 1994
Allobaccha oldroydi Ghorpadé, 1994
Subgenus *Allobaccha* Curran, 1928
Allobaccha (Allobaccha) amphithoe (Walker, 1849)
Allobaccha (Allobaccha) apicalis (Loew, 1858)
Allobaccha (Allobaccha) dispar (Walker, 1859)
Allobaccha (Allobaccha) elegans (Brunetti, 1915)
Allobaccha (Allobaccha) fallax (Austen, 1893)
Allobaccha (Allobaccha) pulchrifrons (Austen, 1893)
Allobaccha (Allobaccha) sapphirina (Wiedemann, 1830)
Allobaccha (Allobaccha) triangulifera (Austen, 1893)
Genus *Allograpta* Osten Sacken, 1875a
Subgenus *Allograpta* Osten Sacken, 1875a
Allograpta (Allograpta) dravida Ghorpadé, 1994
Allograpta (Allograpta) javana (Wiedemann, 1824)
Allograpta (Allograpta) maculipleura (Brunetti, 1913)
Genus *Agnisyrphus* Ghorpadé**, 1994**
Agnisyrphus angara Ghorpadé, 1994
Agnisyrphus brunetti Ghorpadé, 2007
Agnisyrphus gressitti Ghorpadé, 1994
Genus *Asarkina* Macquart, 1842
Asarkina ayyari Ghorpadé, 1994
Asarkina belli Ghorpadé, 1994
Asarkina bhima Ghorpadé, 1994
Asarkina hema Ghorpadé, 1994
Asarkina pitamara Ghorpadé, 1994
Subgenus *Asarkina* Macquart, 1842
Asarkina (Asarkina) africana (Bezzi, 1908)
Asarkina (Asarkina) assimilis (Macquart, 1846)
Asarkina (Asarkina) consequens (Walker, 1856)
Asarkina (Asarkina) ericetorum (Fabricius, 1781)
Asarkina (Asarkina) porcina (Coquillett, 1898)
Genus *Asiobaccha* Violovitsh, 1976
Asiobaccha nubilipennis (Austen, 1893)
Genus *Betasyrphus* Matsumura & Adachi, 1917
Betasyrphus aeneifrons (Brunetti, 1913)
Betasyrphus bazini (Brunetti, 1925)
Betasyrphus fletcheri Ghorpadé, 1994
Betasyrphus isaaci (Bhatia and Shaffi, 1933)
Betasyrphus linga Ghorpadé, 1994
Betasyrphus serarius (Wiedemann, 1830)

Genus *Chrysotoxum* Meigen, 1803

Chrysotoxum antiquum Walker, 1852
Chrysotoxum arcuatum (Linnaeus, 1758)
Chrysotoxum baphyrum Walker, 1849
Chrysotoxum convexum Brunetti, 1915
Chrysotoxum corbetti Ghorpadé, 1994
Chrysotoxum festivum (Linnaeus, 1758)
Chrysotoxum fasciolatum (De Geer, 1776)
Chrysotoxum intermedium Meigen, 1822
Chrysotoxum quadrifasciatum Brunetti, 1923

Genus *Citrogramma* Vockeroth, 1969

Citrogramma amarilla Mengual, 2012
Citrogramma chola Ghorpadé, 1994
Citrogramma citrinum (Brunetti, 1923)
Citrogramma clarum (Hervé-Bazin, 1923)
Citrogramma currani Mengual, 2012
Citrogramma flavigena Wyatt, 1991
Citrogramma frederici Mengual, 2012
Citrogramma marissa Mengual, 2012

Genus *Dasysyrphus* Enderlein, 1938

Dasysyrphus albostriatus (Fallén, 1817)
Dasysyrphus darada Ghorpadé, 1994
Dasysyrphus orsua (Walker, 1852)
Dasysyrphus pandu Ghorpadé, 1994
Dasysyrphus rossi Ghorpadé, 1994

Genus *Didea* Macquart, 1834

Didea fasciata Macquart, 1834
Didea poorva Ghorpadé, 1994
Didea vockerothi Ghorpadé, 1994

Genus *Dideoides* Brunetti, 1908

Dideoides kempi Brunetti, 1923
Dideoides ovatus Brunetti, 1908
Dideoides tigerinus (Bigot, 1885)
Dideoides trilineatus Brunetti, 1923

Genus *Dideopsis* Matsumura & Adachi, 1917

Dideopsis aegrota (Fabricius, 1805)

Genus *Epistrophe* Walker, 1852

Epistrophe carmichaeli Ghorpadé, 1994

Subgenus *Epistrophe* Walker, 1852

Epistrophe (Epistrophe) aequalis (Walker, 1852)
Epistrophe (Epistrophe) griseocinctus (Brunetti, 1923)

Genus *Epistrophella* Dušek and Láska, 1967

Epistrophella shibakawae (Matsumura & Adachi, 1917)

Genus *Episyrphus* Matsumura

Episyrphus (Episyrphus) balteatus (De Geer, 1776)

Episyrphus (Episyrphus) viridaureus (Wiedemann, 1824)
Genus *Eriozona* Schiner, 1860
Subgenus *Eriozona* Schiner, 1860
Eriozona (Eriozona) analis Kertész, 1901
Genus *Eupeodes* Osten Sacken, 1877
Subgenus *Macrosyrphus* Matsumura & Adachi, 1917
Eupeodes (Macrosyrphus) confrater (Wiedemann, 1830)
Subgenus *Metasyrphus* Matsumura in Matsumura & Adachi, 1917
Eupeodes (Metasyrphus) bucculatus (Rondani, 1857)
Eupeodes (Metasyrphus) corollae (Fabricius, 1794)
Eupeodes (Metasyrphus) latifasciatus (Macquart, 1829)
Eupeodes (Metasyrphus) luniger (Meigen, 1822)
Eupeodes (Metasyrphus) nuba (Wiedemann, 1830)
Genus *Ischiodon* Sack, 1913
Ischiodon scutellaris (Fabricius, 1805)
Genus *Leucozona* Schiner, 1860
Subgenus *Ischyrosyrphus* Bigot, 1882b
Leucozona (Ischyrosyrphus) sivae (Bigot, 1882b)
Subgenus *Leucozona* Schiner, 1860
Leucozona (Leucozona) brunetti Ghorpadé, 1994
Leucozona (Leucozona) virendra Ghorpadé, 1994
Genus *Megasyrphus* Dušek & Láska, 1967
Megasyrphus himalayensis Kohli, Kapoor & Gupta, 1988
Eriozona (Megasyrphus) pseudohimalayensis Sengupta, Naskar & Banerjee, 2023b
Genus *Melangyna* Verrall, 1901
Subgenus *Melangyna* Verrall, 1901
Melangyna (Melangyna) remota (Brunetti, 1923)
Genus *Meliscaeva* Frey, 1946
Meliscaeva ceylonica (Keiser, 1958)
Meliscaeva cinctella (Zetterstedt, 1843)
Meliscaeva cinctelloides Ghorpadé, 1994
Meliscaeva darjeelingensis Datta & Chakraborti, 1986
Meliscaeva kusuma Ghorpadé, 1994
Meliscaeva lefroyi Ghorpadé, 1994
Meliscaeva magnifica Ghorpadé, 1994
Meliscaeva mathisi Ghorpadé, 1994
Meliscaeva tribeni (Nayar, 1968a, b)
Genus *Paragus* Latreille, 1804
Subgenus *Paragus* Latreille, 1804
Paragus (Paragus) ambalaensis Sodhi & Singh, 1991
Paragus (Paragus) annandalei Ghorpadé, 1992
Paragus (Paragus) bicolor (Fabricius, 1794)
Subgenus *Serratoparagus* Vujić et al. 2008
Paragus (Serratoparagus) auritus Stuckenberg, 1954b

Paragus (Serratoparagus) crenulatus Thomson, 1869
Paragus (Serratoparagus) serratus (Fabricius, 1805)
Paragus (Serratoparagus) yerburiensis Stuckenberg, 1954b

Subgenus *Pandasyopthalmus* Stuckenberg, 1954b
Paragus (Pandasyopthalmus) atratus Meijere, 1906
Paragus (Pandasyopthalmus) haemorrhous Meigen, 1822
Paragus (Pandasyopthalmus) politus Wiedemann, 1830
Paragus (Pandasyopthalmus) rufocinctus (Brunetti, 1908)
Paragus (Pandasyopthalmus) tibialis (Fallén, 1817)

Genus *Parasyrphus* Matsumura & Adachi, 1917
Parasyrphus aeneostoma (Matsumura & Adachi, 1917)
Parasyrphus kashmiricus Ghorpadé, 1994
Parasyrphus thompsoni Ghorpadé, 1994

Genus *Rhinobaccha* Meijere, 1908
Rhinobaccha krishna Ghorpadé, 1994
Rhinobaccha peterseni Ghorpadé, 1994

Genus *Scaeva* Fabricius, 1805
Scaeva albomaculata (Macquart, 1842)
Scaeva latimaculata (Brunetti, 1923)
Scaeva opimius (Walker, 1852)
Scaeva pyrastri (Linnaeus, 1758)
Scaeva selenitica (Meigen, 1822)

Genus *Sphaerophoria* Le Peletier & Serville, 1828
Sphaerophoria knutsoni Ghorpadé, 1994
Sphaerophoria ladaskhensis Ghorpadé, 1994
Sphaerophoria vockerothi Joseph, 1970

Subgenus *Knutsoniana* Baralov & Mutin, 1977
Sphaerophoria (Knutsoniana) assamensis Joseph, 1970
Sphaerophoria (Knutsoniana) viridaenea Brunetti, 1915

Subgenus *Sphaerophoria* Le Peletier & Serville, 1828
Sphaerophoria (Sphaerophoria) bengalensis Macquart, 1842
Sphaerophoria (Sphaerophoria) indiana Bigot, 1884b
Sphaerophoria (Sphaerophoria) macrogaster (Thomson, 1869)
Sphaerophoria (Sphaerophoria) rueppellii (Wiedemann, 1830)
Sphaerophoria (Sphaerophoria) scripta (Linnaeus, 1758)

Genus *Syrphus* Fabricius, 1775
Syrphus dalhousiae Ghorpadé, 1994
Syrphus howletti Ghorpadé, 1994

Subgenus *Syrphus* Fabricius, 1775
Syrphus (Syrphus) fulvifacies Brunetti, 1913
Syrphus (Syrphus) ribesii (Linnaeus, 1758)
Syrphus (Syrphus) torvus Osten Sacken, 1875b
Syrphus (Syrphus) vitripennis Meigen, 1822

2.6 Discussion

Hoverflies represent over 6300 global records (Skevington et al. 2019), of which India accounts for 368 species belonging to 77 genera under four subfamilies (Thompson and Ghorpadé 1998; Sengupta et al. 2016a, 2024). Among these subfamilies, Eristalinae harbors the highest number of species (203), followed by Syrphinae (147), Microdontinae (17), and Pipizinae (1) (Fig. 2.10). Upon considering Tribe-wise distribution of species, Tribe Syrphini contributes the highest proportion of species (35.59%), followed by Milesiini (17.39%) and Eristalini (13.04%) (Fig. 2.11). Among the genera, *Eumerus* and *Eristalinus* (both belonging to Eristalinae) are the most speciose, represented by 17 and 14 species; *Milesia*, *Monoceromyia*, and *Sphegina* have 13 species. Notably, the subfamily Eristalinae is the most widespread and species-rich taxa not only in India but also globally (Evenhuis and Pape 2024). The remarkable ecological adaptability of Eristalinae, especially due to their diverse larval feeding strategies, has enabled members of this subfamily to occupy nearly every conceivable habitat worldwide, contributing to its dominance within Syrphidae.

In India, hoverfly research has traditionally emphasized adult taxonomy and morphological descriptions, with sporadic attention given to aphidophagous species in the context of biological control. However, the ecological significance of Syrphidae extends well beyond pollination and aphidophagy. Many essential ecosystem services, particularly those mediated by the diverse larval feeding strategies

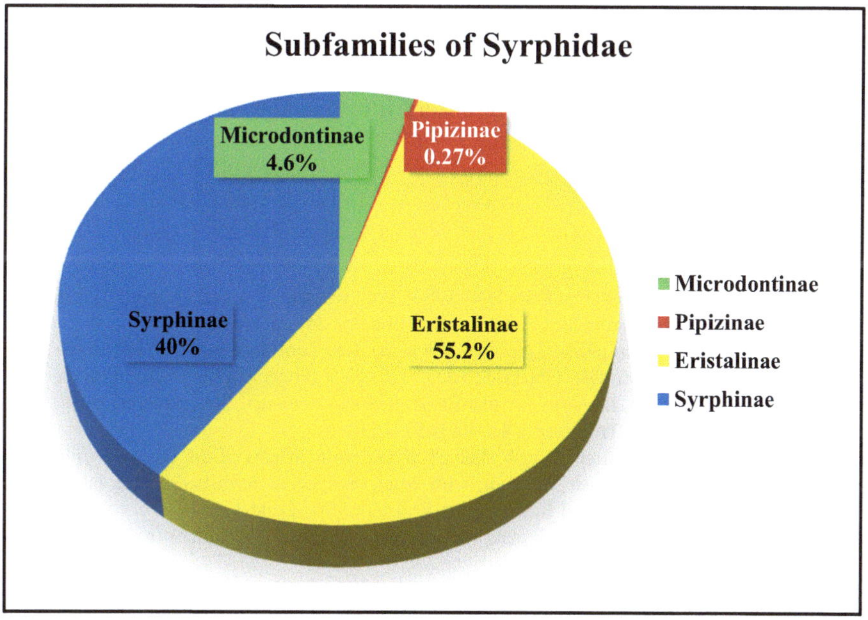

Fig. 2.10 Percentage of species in each subfamily of Syrphidae

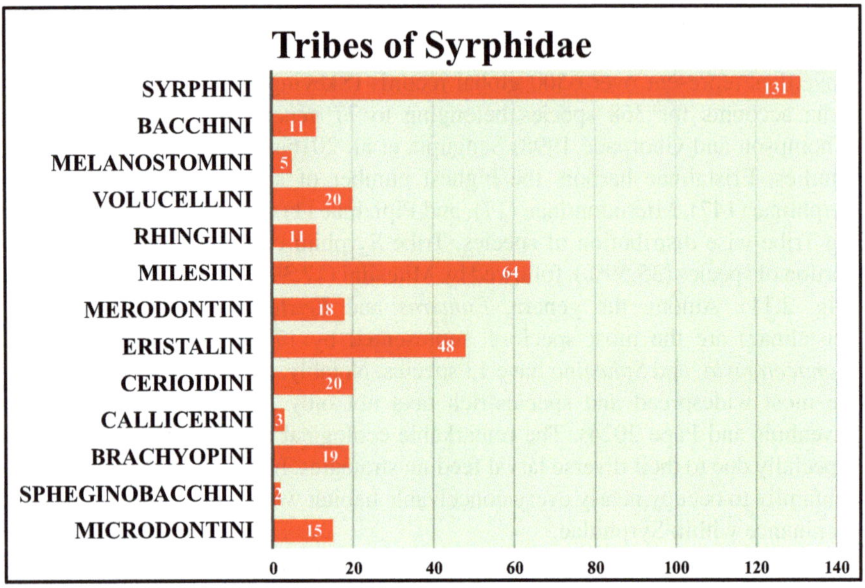

Fig. 2.11 Number of species in each tribe of Syrphidae

of Eristalinae—ranging from aphid predation to saprophagy in organic pollutant-rich aquatic habitats—remain poorly documented in the Indian context. Studies on larval biology, life history traits, and functional ecology are especially scarce, creating a substantial gap in the understanding of how these taxa contribute to decomposition, nutrient cycling, and ecosystem resilience. To bridge this knowledge gap, comprehensive field surveys combined with integrative approaches, including molecular systematics, detailed life-history investigations, and experimental ecological studies, are urgently needed.

References

Agarwala BK, Láska P, Roychaudhuri DN (1984) Prey records of aphidophagous syrphid flies from India (Diptera: Syrphidae). Acta Entomol Bohemoslov 81:15–21

Anand D (1986) Studies on Indian Syrphidae (Diptera). Rec Zool Surv India 83(1–4):1–76

Anooj SS, Kalia V, Krishna GK, Ghorpadé KD (2020) New biogeographic distribution record of phytophagous syrphid, *Eumerus vestitus* Bezzi, its biosystematics, host preferences and association behaviour. Int J Trop Insect Sci 40(3):527–538

Austen EE (1893) Descriptions of new species of dipterous insects of the family Syrphidae in the collection of the British Museum, with notes on species described by the late Francis Walker. –Part I. Bacchini and Brachyopini. Proc Zool Soc Lond 132–164

Baralov GD, Mutin VA (1977) New data on hoverflies (Diptera, Syrphidae) from the Far East. Zoologicheskii Zhurnal 56(5):731–740

Barkalov AV (2002) A subgeneric classification of the genus *Cheilosia* Meigen, 1822 (Diptera, Syrphidae). Entomologicheskoe Obozrenie 81:218–234

Barkalov AV (2012) Subgeneric classification of the genus *Sphaerophoria* Lep. et Serv. (Diptera, Syrphidae). Euroasian Entomol J 11:285–290

Becker T (1894) Revision der Gattung *Cheilosia* Meigen. Nova acta Academiae Caesareae Leopoldino-Carolinae Germanicae Naturae Curiosorum 62(3):194–521
Bezzi M (1908) Secondo contributo alla conoscenza del genere Asarcina. Ann Hist Nat Mus Natl Hung 6:495–504
Bezzi M (1912) Ditteri raccolti da Leonardo Fea durante il suo viaggio nell'Africa occidentale. Parte 1a: Syrphidae. Annali del Museo civico di Storia naturale di Genova 45:400–443
Bhatia HL, Shaffi M (1933) Life-histories of some Indian Syrphidae. Indian J Agric Sci 2(6):543–570
Bigot JMF (1880) Dipteres nouveaux ou peu connus. 13e partie. XX. Quelques Dipteres de Perse et du Caucase. Ann Soc Entomol Fr 5(10):139–154
Bigot JMF (1882a) Descriptions de quatre genres nouveaux de la tribu des syrphides (Syrphidae auctorum), ainsi que celles de deux nouvelles espèces. Bulletin des Séances [Bimensuel] de la Société Entomologique de France 6:78–79
Bigot JMF (1882b) Diagnoses de genres et espèces inedits de syrphides. (4e et derniere partie). Bulletin Bimensuel de la Société Entomologique de France 15:173
Bigot JMF (1884a) Diptères nouveaux ou peu connus. 22e partie, XXXII: Syrphidi (2e). espèces nouvelles, No 1er. Annales de la Société Entomologique de France 3(6):315–356
Bigot JMF (1884b) Diptères nouveaux ou peu connus. 24 partie, XXXII: Syrphidi (2 partie). espèces nouvelles, no. III. Annales de la Société Entomologique de France Series 6, 4, 73–80, 81–116.
Bigot JMF (1885) Diptères nouveaux ou peu connus. 28e partie. XXXVI: Syrphidi. Addenda au memoire publie dans les Annales de la Société entomologique de France (Annees 1883–1884). Nouvelles espèces. Annales de la Société Entomologique de France Series 6(5):247–252
Biswas S, Lahiri AR, Ghosh AK (1975) A preliminary study of the Insect fauna of Meghalaya
Brunetti E (1907) Notes on the Oriental Syrphidae. Part I. Rec Indian Mus 1:379–380
Brunetti E (1908) Notes on Oriental Syrphidae with descriptions of new species. Part I. Rec Indian Mus 2:49–96
Brunetti E (1913) New and interesting Diptera from the eastern Himalayas. Rec Indian Mus 9(5):255–277
Brunetti E (1915) Notes on Oriental Syrphidae: with descriptions of new species. Part II. Rec Indian Mus 11:201–256
Brunetti E (1917) Diptera of the Simla District. Rec Indian Mus 13:59–101
Brunetti E (1923) The Fauna of British India, including Ceylon and Burma. Diptera. Volume 3. by hoverflies in the Anthropocene. Proc R Soc B Biol Sci 287(1927):20200508
Brunetti E (1924) *Microdon Apiformis* Brun., Renamed. Rec Zool Surv India 26(2):153
Brunetti E (1925) Some notes on Indian Syrphidae, Conopidae, and Oestridae. Rec Indian Museum 27:75–79
Campoy A, Pérez-Bañón C, Nielsen TR (2017) Micromorphology of egg and larva of Eristalis fratercula, with an updated key of Eristalis species with known third instar larvae (Diptera: Syrphidae). Acta Entomologica Musei Nationalis Pragae 57(1):215–227
Campoy A, Pérez-Bañón C, Aznar D, Rojo S (2020) Description of the preimaginal stages of the golden native dronefly from Australia, *Eristalinus punctulatus* (Macquart, 1847) (Diptera: Syrphidae). Aust Entomol 59(4):784–793
Coe RL (1964) Diptera from Nepal. Syrphidae. Bull Br Mus Nat Hist 15:255–290
Coquillett DW (1898) Report on a collection of Japanese Diptera, presented to the U.S. National Museum by the Imperial University of Tokyo. Proc USA Natl Museum 21(1146):301–340
Curran CH (1925) Contribution to a monograph of the American Syrphidae north of Mexico. Kansas Univ Sci Bullet 15(1):7–216
Curran CH (1928) The Syrphidae of the Malay Peninsula. J Fed Malay States Mus 14:141–324
Curran CH (1929) New Syrphidae and Tachinidae. Ann Entomol Soc Am 22:489–510
Curran CH (1931) Records and descriptions of Syrphidae from North Borneo including Mt Kinabalu. J Federal Malay States Museums 16:333–376
Curran CH (1933) Three new Diptera from India. Stylops 2:45–48

Datta M, Chakraborti, M (1986) New records of Syrphidae from Darjeeling (West Bengal) and Sikkim, India, with description of *Meliscaeva darjeelingensis* spec. nov. (Diptera). Opusc Zool Flumin 6:1–19

De Geer C (1776) Mémoires pour servir à l'histoire des insectes. Tome Sixième. P. Hesselberg, Stockholm, viii + 523 pp

De Meijere JCH (1908) Neue Beiträge zur Kenntnis der Dipteren von Sumatra. Tijdschrift voor Entomologie 51:118–196

De Meijere JCH (1910) Studien über südostasiatische Dipteren. Tijdschrift voor Entomologie 53:231–268

De Meijere JCH (1914) Studien über südostasiatische Dipteren. Tijdschrift voor Entomologie 57:1–64

Deoras PJ (1943) Description of and biological notes on a new species of Syrphidae from India. Diptera, Syrphidae: eleven new records and notes on other species. Proc Zool Soc 27:23–27

der Sluijs V, Jeroen P et al (2013) Neonicotinoids, bee disorders and the sustainability of pollinator services. Curr Opin Environ Sustain 5:293

van Doesburg PH (1955) Report on the syrphid flies, collected by the "Fourth Dutch Karakorum Expedition, 1935 (Mededelingen over Syrphidae XIII). Beaufortia 5:47–51

Doyle T, Hawkes WLS, Massy R, Powney GD, Menz MHM, Wotton KR (2020) Pollination by hoverflies in the Anthropocene. Proc R Soc B Biol Sci 287(1927):20200508. https://doi.org/10.1098/rspb.2020.0508

Dušek J, Láska P (1967) Versuch zum Aufbau eines naturlichen Systems mitteleuropäischer Arten der Unterfamilie Syrphinae (Diptera), [A study of the structure of a natural classification of central European species of the subfamily Syrphinae (Diptera)]. Acta Scientarum Naturalium Brno 1:349–390

Edwards FW (1915) [Title]. [Journal/Monograph], Volume(Issue), pages

Enderlein G (1938) Beiträge zur Kenntnis der Syrphiden. Sitzungsberichte der Gesellschaft Naturforschender Freunde Berlin 1937:192–237

Evenhuis NL, Pape T (2024) Systema Dipterorum, Version 5.0. Accessed on 2 Feb 2024. Available at http://www.diptera.org/. Catalogue of Life, part III. Br Mus Nat Hist Lond 3:485–687

Fabricius JC (1775) Systema entomologiae, sistens insectorvm classes, ordines, genera, species, adiectis synonymis, locis, descriptionibvs, observationibvs. Flensburgi et Lipsiae, Korte, p 832

Fabricius JC (1781) Species insectorum, exhibentes eorum differentias specificas, synonyma auctorum, locanatalia, metamorphosin, adjectis observationibus, descriptionibus, vol 1. Impensis CE Bohnii

Fabricius JC (1787) Mantissa Insectorum sistens species nuper detectas, vol 2. C.G. Proft, Hafniae [= Copenhagen], 382 pp

Fabricius JC (1794) Entomologia systematica emendata et aucta. Secundum classes, ordines, genera, species adjectis synonimis, locis, observationibus, descriptionibus. Tom. IV. CG Proft, Fil et Soc, Hafniae [= Copenhagen]. 4:472

Fabricius JC (1805) Systema antliatorum secundum ordines, genera, species adiecta synony- mis, locis, observationibus, descriptionibus. C. Reichard, Brunsvigae [= Brunswick], xiv +

Fallén FC (1817) Syrphici Sveciae. In: Diptera Sveciae/descripta a Carlo Frederico Fallén. Literis Berlingianis, Lundae, pp 23–62

Fluke CL Jr (1943) A new genus and new species of Syrphidae (Diptera) from Ecuador. Ann Entomol Soc Am 36:425–431

Frey R (1946) Ubersicht der Gattungen der Syrphiden-Unterfamilie Syrphinae (Syrphine + Bacchinae). Notulae Entomologicae 25:152–172

Garibaldi LA, Steffan-Dewenter I, Kremen C, Morales JM, Bommarco R, Cunningham SA, Holzschuh A (2011) Stability of pollination services decreases with isolation from natural areas despite honey bee visits. Ecol Lett 14(10):1062–1072

Garibaldi LA, Steffan-Dewenter I, Winfree R, Aizen MA, Bommarco R, Cunningham SA, Bartomeus I (2013) Wild pollinators enhance fruit set of crops regardless of honey bee abundance. Science 339(6127):1608–1611

Geoffroy EL (1762) Histoire abregée des insects qui se trouvent aux environs de Paris: dans laquelle ces animaux sont rangés suivant un ordre méthodique, vol 2. Durand, Paris, p 690
Ghahari H, Hayat R, Tabari M, Ostovan H (2008) Hover flies (Diptera: Syrphidae) from rice fields and around grasslands of Northern Iran. Mun Entomol Zool 3(1):275–284
Ghorpadé KD (1981) Taxonomic revision of Syrphidae (Diptera: Syrphidae) from the Indian subcontinent. Bangalore, University of Agricultural Sciences
Ghorpade KD (1982) A new Callicera (Diptera: Syrphidae) from the northwest Himalaya
Ghorpadé K (1994) Diagnostic keys to new and known genera and species of Indian subcontinent Syrphini (Diptera: Syrphidae). Colemania: Insect Biosyst 3:1–15
Ghorpadé K (2007) The genus *Agnisyrphus* Ghorpadé (Diptera-Syrphidae), peculiar to the Oriental Region, with notes on phylogeny, evohistory and panbiogeography. Colemania, 3(14):1–35
Ghorpadé K (2009) Some nomenclatural notes on Indian subregion Syrphini (Diptera—Syrphidae). Colemania 15:3–13
Ghorpadé K (2012) Notes on nomenclature, taxonomy and phylogeny of the genus Chrysotoxum Meigen (Diptera—Syrphidae) in the Oriental region. Colemania 32:1–4. mapress.com
Ghorpadé K (2014) An updated check-list of the Hover-flies (Diptera–Syrphidae) recorded in the Indian subcontinent. Colemania 44:1–30
Gilbert FS (1985) Morphometric patterns in hoverflies (Diptera, Syrphidae). Proc R Soc Lond B 224:79–90
Gilbert FS, Harding EF, Line JM, Perry I (1985) Morphological approaches to community structure in hoverflies (Diptera, Syrphidae). Proc R Soc Lond B Biol Sci 224(1234):115–130
Greene CT (1949) Five new species of *Graptomyza* (Diptera, Syrphidae). Proc Entomol Soc Wash 51:77–80
Guérin FÉ (1833) Notice sur les métamorphoses des Cératopogons et description de deux espèces nouvelles de ce genre, découvertes aux environs de Paris. Annales de la Société Entomologique de France, 2, 161–167. Magnolia Press+2zobodat.at+2
Harris M (1776–1780) An exposition of English insects, with curious observations and remarks, wherein each insect is particularly described; its parts and properties considered; the different sexes distinguished, and the natural history faithfully related. The whole illustrated with copper plates, drawn, engraved, and coloured, by the author. Printed for the Author, London, 166 + [4] pp., pls. 1–50
Heo CC, Rahimi R, Mengual X et al (2019) *Eristalinus arvorum* (Fabricius, 1787) (Diptera: Syrphidae) in human skull: a new fly species of forensic importance. J Forensic Sci 65:276
Hervé-Bazin J (1914) Note sur quelques Syrphides (Diptera) provenant de Java et de l'Inde. Avec ladescription d'un genre nouveau. Insecta, Revue Illustr Entomol 41:149–154, 6 figs
Hervé-Bazin J (1922) Description d'une nouvelle espèce de Korinchia de l'Inde (Dipt. Syrphidae). Bulletin de la Société Entomologique de France 27(9):122–124
Hervé-Bazin J (1923) Etudes sur les "*Lathyrophthalmus*" (Diptera, Syrphidae) d'extreme-Orient
Hippa H (1974) On the taxonomy of the Ceylonese and southern Indian species of the genus Xylota Meigen (Dipt, Syrphidae). Ann Entomol Fenn 40(2):56–60, 10 figs
Hippa H (1978) Classification of Xylotini (Diptera, Syrphidae). Acta Zool Fenn 156:1–153
Hippa H (1985) Infrageneric classification and revision of the Oriental species of the genus *Brachypalpoides* Hippa (Diptera, Syrphidae). Acta Entomol Fennica 45:1–20
Hippa H (1990) The genus *Milesia* Latreille (Diptera, Syrphidae). Acta Zool Fenn 187:1–226
Hippa H, Van JS, Mutin VA (2015) The genus Sphegina Meigen (Diptera, Syrphidae) in a biodiversity hotspot: the thirty-six sympatric species in Kambaiti, Myanmar. Zootaxa 3954:1–67
Holzschuh A (2011) Stability of pollination services decreases with isolation from natural areas despite honey bee visits. Ecol Lett 14(10):1062–1072
Hull FM (1941) Descriptions of some new species of Syrphidae. Psyche 48(4):149–165
Hull FM (1942) The genus Ferdinandea Rondani. J Wash Acad Sci 32(8):239–241
Hull FM (1944) Some Syrphid fly Genera. Entomol News 55:203–205
Hull FM (1949) The morphology and inter-relationship of the genera of syrphid flies, recent and fossil. Transactions of the Zoological Society of London 26:257–408

Hull FM (1950) Studies upon Syrphid flies in the British Museum (natural history). Ann Mag Nat Hist 3(12):603–624

Ilse D (1949) Colour discrimination in the dronefly, Eristalis tenax. Nature 163:255–256

Jaennicke F (1867) Neue exotische Dipteren. Abhandlungen, herausgegeben von der

Joseph ANT (1970) New and little-known species of Syrphidae (Diptera) from India. Orient Insects 4(1):1–16

Joseph ANT (1970c) Two new and two known species of *Sphaerophoria* St. Fargeau and Serville, 1828 (Dipt.Syrphidae). Eos 45:165–172

Joseph ANT, Parui P (1972) Diptera from NEFA and Assam foothills part II. Kameng Frontier Division, Subansiri Frontier Division, North Lakhimpur and Assam foot hills. Rec Zool Surv India 66:103–128

Joseph ANT, Parui P (1973) Diptera from NEFA and Assam foothills part III. Siang Frontier Division and North Lakhimpur. Rec Zool Surv India 67:325–342

Joseph ANT, Parui P (1976) Diptera from Arunachal Pradesh and Assam foothills, part IV. Lohit District, Lakhimpur District and Sibsagar District. Rec Zool Surv India 69:275–290

Joseph ANT, Parui P (1977a) On a small collection of Diptera from Chota Nagpur, Bihar. Rec Zool Surv India 72:227–238

Joseph ANT, Parui P (1977b) Diptera from Arunachal Pradesh and Assam foothills, part V. Tirap Division. Rec Zool Surv India 72:333–339

Joseph ANT, Parui P (1986) Diptera from Silent Valley. Rec Zool Surv India 84:157–164

Joseph ANT, Sharma HS (1976) News records of Syrphidae (Diptera) from Sikkim, India. Newsl Zool Surv India 2(6):252–254

Joshi S, Ballal CR (2013) Syrphid predators for biological control of aphids. J Biol Control 27(3):151–170

Kanervo E (1938) Zur Systematik und Phylogenie der westpalaarktischen Eristalis Arten (Dipt. Syrphidae) mit einer Revision derjenigen Finnlands. Annales Universitasis Turkuensis (A) 6(4):54

Kar O, Ghosh D, Mukherjee A, Mukherjee K, Pramanik D, Sarkar S, Sengupta J, Naskar A, Banerjee D (2024) Molecular identification of ecologically relevant hoverflies (Diptera, Syrphidae) from eastern India. Entomon 49(3):295–308

Keiser F (1958) Beitrag zur Kenntnis der Syrphidenfauna von Ceylon (Dipt.). Rev Suisse Zool 65:185–239

Kertész K (1901) Neue und bekannte Dipteren in der sammlungdes Ungarischen National-Museums. Természetrajzi Füuzetek, tab XX 24:405–432

Khajuria M, Tripathi NK, Bhatti AA (2015) Karyotypicand morphometric analysis of an Aphidophagous Hoverfly, Episyrphus balteatus (Diptera: Syrphidae) from Jammu Region (J&K), India. Int J Adv Res 3:194–198

Khatua P, Mohapatra LN, Nayak P, Dibyarani (2020) Feeding potential and functional response of syrphid fly, Eumerus albifrons Walker to cowpea aphid, Aphis craccivora Koch. J Entomol Zool Stud 8(5):2324–2327

Klecka J, Hadrava J, Biella P, Akter A (2018) Flower visitation by hoverflies (Diptera: Syrphidae) in a temperate plant-pollinator network. PeerJ 6:e6025

Kluser S, Peduzzi P (2007) Global pollinator decline: a literature review. UNEP/GRIDEurope

Knutson LV, Thompson FC, Vockeroth JR (1975) Family Syrphidae. In: Delfinado MD, Hardy DE (eds) Catalog of the diptera of the oriental region 2. University Press of Hawaii, Honolulu, pp 307–374

Kohli VK, Kapoor VC, Gupta SK (1988) On one new genus and nine species of syrphid flies (Diptera: Syrphidae) from India. J Insect Sci 1:113–127

Korzets Z, Bernheim J, Lengy J, Gold D (1993) Human urogenital myiasis due to Eristalis larva: an unusual cause of ureteric obstruction. Nephrol Dial Transplant 8:874–876

Lambeck HJP, van Brink JM (1973) Contribution to the knowledge of the taxonomy, faunal composition and cytology of the syrphid flies (Diptera: Syrphidae) of Kashmir (India). I. Taxonomic account and faunal composition. Genen Phaenen 16(3):87–100, 2 figs

Latreille PA (1804) Tableau méthodique des insectes. Société de Naturalistes et d'Agriculteurs (Ed.). Nouvelle dictionnaire d'Histoire naturelle 24(3):129–200

Le Peletier ALM, Serville JGA (1828) Various articles. In: Latreille PA, Lepeletier ALM, Serville JGA, Guérin-Méneville FE (eds) Entomologie, ou histoire naturelle des Crustaces, des Arachnides et des insectes. Encyclopédie Méthodique, Histoire Naturelle. Insectes, vol 10. Agasse, Paris, pp 345–833

Lima MAP, Cutler GC, Mazzeo G, Hrncir M (2022) Editorial: The decline of wild bees – causes and consequences. Front Ecol Evol. https://doi.org/10.3389/fevo.2022.1027169

Linnaeus C (1758) Systema Naturae per Regna Tria Naturae, Secundum Classes, Ordines, Genera, Species, cum Characteribus, Differentiis, Synonymis, Locis, vol 1, 10th edn. Laurentii Salvii, Holmiae

Linnaeus C (1761) Fauna Suecica, 2nd edn. Laurentii Salvii, Stockholm

Loew H (1858) Bidrag till kännedomen om Afrikas Diptera (part). Öfvers K Vet Akad Förh 14(9):337–383

Lunau K (2014) Visual ecology of flies with particular reference to colour vision and colour preferences. J Comp Physiol A 200:497

Lyneborg L, Barkemeyer W (2005) The genus *Syritta*. A world revision of the genus *Syritta* Le Peletier & Serville, 1828 (Diptera: Syrphidae). Entomon 15:224

Macquart PJM (1829) Insectes dipteres du nord de la France. Syrphies, Lille 1827:12–23

Macquart PJM (1834) Histoire naturelle des insectes. Diptères. Tome premiere. Roret, Paris. 578 + 8 pp., 12 pls

Macquart PJM (1842) Diptères exotiques nouveaux ou peu connus. Tome deuxieme. 2e partie. Memoires de la Société Royale des Sciences de l'Agriculture et des Arts, Lille, 1, 65–200.

Macquart J (1846) Diptères exotiques nouveaux ou peu connus. Supplément [I]. Memoires de la Société (Royale) des Sciences de l'Agriculture et des Arts de Lille, 1844(1845):133–364.

Macquart PJM (1850) Dipteres exotiques nouveaux ou peu connus. 4e supplement. Memoirs de la Societe Royale des Sciente, de l'Agriculture et des Art, de Lille 1849:309–479

Macquart PJM (1855) Diptères exotiques nouveaux ou peu connus. 5.e supplément. Mémoires de la Société des Sciences, Agriculture et des Arts, de Lille Series 2, 1:25–156

Maqbool A, Wachkoo AA, Yatoo SF, Skevington JH, van Steenis J (2024) A new species of the genus Spilomyia Meigen, 1803 (Diptera: Syrphidae) from South Asia. J Asia Pac Entomol 27(2):102254

Matsumura S, Adachi J (1917) Synopsis of the economic Syrphidae of Japan (pt. III). Entomol Magaz Kyoto 3:14–46

Mengual X, Barkalov AV (2019) Two new species of *Rohdendorfia* (Diptera: Syrphidae) from Central Asia. Acta Entomologica Musei Nationalis Pragae 59(1):325–336

Megual X, Barkalov AV (2023) A new species of *Cheilosia* Meigen (Diptera: Syrphidae) from high mountains of India. Studia Dipterologica Suppl 23:227–235

Meigen JW (1803) Versuch Einer Neuen Gattungeintheilung der Europaischen Zweiflugeligen Insekten. Magazin fur Insektenkunde, Herausgegeben von Karl Illiger 2:259–281

Meigen JW (1822) Systematische Beschreibung der bekannten europäischen zweiflügeligen Insekten. Dritter Theil, Schulz-Wundermann, Hamm, Germany, 416 pp.

Meijere JCH de (1906) Diptera. Résultats de l'Expédition Scientifique Néerlandaise à la Nouvelle-Guinée en 1903 sous les auspices de Arthur Wichmann. Nova Guinea 5(1):67–99

Mengual X [with inputs by K. Ghorpadé and F.C. Thompson] (2012) The flower fly genus Citrogramma Vockeroth (Diptera, Syrphidae), illustrated revision with descriptions of new species. Zool J Linnean Soc 164:99–172

Mik JL (1897) Einige Bemerkungen zur Dipteren–Familie der Syrphiden. Wiener Ent Zeit 1:113–119

Miličić M, Vujić A, Jurca T, Cardoso P (2017) Designating conservation priorities for Southeast European hoverflies (Diptera: Syrphidae) based on species distribution models and species vulnerability. Insect Conserv Divers 10(4):354–366

Mitra B, Parui P, Sharma RM (2004) A preliminary study on the Dipteran Flower visitors/pollinators of Himachal Pradesh. Ann For Res 12(1):119–124

Mitra B, Mukherjee M, Banerjee D (2008a) A check-list of hover-flies (Diptera Syrphidae) of Eastern Hinlalayas. Rec Zool Surv India Occ Paper No 284:1–47
Mitra B, Parui P, Mukherjee M, Sharma RM (2008b) Fauna of Pin Valley National Park. Conserv Area Ser 34:75–84
Mitra B, Roy S, Imam I, Ghosh M (2015) A review of the hover flies (Syrphidae: Diptera) from India. Int J Fauna Biol Stud 2:61–73
Mukherjee M, Parui P, Mitra B (2006) Diptera: Syrphidae. Fauna of Nagaland, State Fauna ser, Zoological Survey of India, Kolkata. 12:185–192
Mutin VA (1998) Four new species of the genus Sphegina Meigen, 1822 (Diptera: Syrphidae) from Russia and India. Int J Dipterol Res 9(3):237–241
Nayar JL (1968a) A contribution to our knowledge of high altitude Syrphidae (Cyclorrhapha: Diptera) from N.W. Himalaya. Part I—subfamily Syrphinae. Agra Univ J Res: Sci 16(3):121–131
Nayar JL (1968b) A contribution to our knowledge of high altitude Syrphidae (Cyclorrhapha: Diptera) from N.W. Himalaya. Part II—subfamily Eristalinae. Agra Univ J Res: Sci 16(3):27–31
Newman E (1838) Entomological Notes. Entomol Magaz 5:372–402
Nielsen TR (2004) A new *Platycheirus* of the *ambiguus* group (Diptera, Syrphidae) from Kashmir. Volucella 7:31–34
Nielsen TR (2007) *Platycheirus cryophilus* spec. nov. (Diptera, Syrphidae) from Kashmir. Volucella 8:95–97
Osten Sacken CR (1875a) A list of the North American Syrphidae. Bullet Buffalo Soc Nat Sci 3:38–71
Osten Sacken CR (1875b) On the North American species of the genus *Syrphus* (in the narrowest sense). Proc Boston Soc Nat Hist (1875–1876) 18:135–153
Osten Sacken CR (1877) Western Diptera: descriptions of new genera and species of Diptera from the region west of the Mississippi and especially from California. Bull US Geol Geogr Surv Terr 3:189–354
Panzer GWF (1806) Favnae insectorvm germanicae initia oder Devtschlands Insecten. H. 103. Felsecker, Nurnberg [= Nuremberg], 24 pp., 24 pls
Paramonov SJ (1927) Dipterologische Fragmente. V–VII. *Zbirnyk Prats Zoologichnogo Muzeyu, Kyiv* (= Travaux du Musée Zoologique, Kyiv) 2: 73–81. [= Trudy Fizychno-Matematychnogo Viddilu Ukrainskoi Akademii Nauk (= Mémoires de la Classe des Sciences Physiques et Mathématiques du Académie des Sciences de l'Ukraïne) 4(4): 317–325]
Pérez-Bañón C, Rojo S, Ståhls G, Marcos-García M (2003a) Taxonomy of European *Eristalinus* (Diptera: Syrphidae) based on larval morphology and molecular data. Eur J Entomol 100:417–428
Pérez-Bañón C, Rotheray G, Hancock G, Marcos-García MA, Zumbado MA (2003b) Immature stages and breeding sites of some Neotropical saprophagous syrphids (Diptera: Syrphidae). Ann Entomol Soc Am 96:458–471. https://doi.org/10.1603/0013-8746(2003)096[0458:ISABSO]2.0.CO;2
Pérez-Bañón C, Hurtado P, García-Gras E, Rojo S (2013) SEM studies on immature stages of the drone flies (Diptera, Syrphidae): *Eristalis similis* (Fallen, 1817) and *Eristalis tenax* (Linnaeus, 1758). Microsc Res Tech 76:853–861
Pérez-Bañón C, Rojas C, Vargas M, Mengual X, Rojo S (2020) A world review of reported myiases caused by flower flies (Diptera: Syrphidae), including the first case of human myiasis from *Palpada scutellaris* (Fabricius, 1805). Parasitol Res 119:815
Potts SG, Biesmeijer JC, Kremen C, Neumann P, Schweiger O, Kunin WE (2010) Global pollinator declines: trends, impacts and drivers. Trends Ecol Evol 25(6):345–353
Rafinesque CS (1815) Analyse de la nature, ou tableau de l'univers et des corps organisés. Published by the author, Palermo, p 224
Reemer M, Ståhls G (2013) Generic revision and species classification of the Microdontinae (Diptera, Syrphidae). ZooKeys 288:1–213
Rondani C (1844) Proposta della formazione di un genere nuovo per due specie di insetti ditteri. Memoria nona per servir alla ditterologia italiana. Nuovi annali delle scienze naturali Bologna 2(2):193–202

Rondani C (1845) Ordinamento sistematico dei generi italiani degli insetti ditteri. Nuovi Annali delle Scienze Naturali e Rendiconto delle Sessioni della Società Agraria e dell'Accademia delle Scienze dell'Istituto di Bologna 1844, Series 2, 2: 443–459

Rondani C (1850) Nota sexta pro dipterologia italicade nova specie generis Ceriae Fabricii. Annales de la Sociétě Entomologique de France 8(2):211–214

Rondani C (1856) Dipterologiae Italicae Prodromus. Vol: I. Genera italica ordinis Dipterorum ordinatim disposita et distincta et in familias et stirpes aggregata. A. Stocchi, Parmae [= Parma], 226 + [2] pp

Rondani C (1857) Species italicae ordinis dipterorum in genera characteribus definita, ordinatim collectae, methodo analatica distinctae, et novis vel minus cognitis descriptis. Pars tertia: Muscidae, Tachininarum complementum. Dipt Ital Prod 2:134–135

Rondani C (1865) Diptera Italica non vel minus cognita descripta vel annotata observationibus nonnullis additis. Fasc. I. Oestridae-Syrphidae-Conopidae; Fasc. II. Muscidae. Atti Soc Ital Sci Nat Mus Civ Stor Nat Milano 8:127–146

Rossi P (1794) Mantissa Insectorum, exhibens species nuper in Etruria collectas, adjectis Faunae Etruscae illustrationibus ac emendationibus. Tomus secundus. Polloni, Pisa, p 154

Rotheray GE, Gilbert F (1999) Phylogeny of Palaearctic Syrphidae (Diptera): evidence from larval stages. Zool J Linnean Soc 117:1–122

Sack P (1913) H. Sauter's Formosa-Ausbeute. Syrphiden I. (Dipt.). Entomologische Mitteilungen 2(1):1–10

Sack P (1928) Appendix. pp. In: O.W. Richards, Bombus and Volucella in the Himalayas

Sankararaman H, Johnson AD, Manickavasagam S, Pennards G (2020) First record of two interesting genera of hover flies (Diptera: Syrphidae) in South India. J Insect Biodivers 14(2):54–63

Sankararaman H, Anooj SS, Mengual X (2022) Review of Indian species of Monoceromyia Shannon (Diptera: Syrphidae) with description of two new species. J Asia Pac Entomol 25:101820

Saunders WW (1842) Descriptions of four new dipterous insects from Central and Northern India. Trans Entomol Soc London [1841] 3(1):59–81

Schiner IR (1860) Vorlaufiger Commentar zum dipterologischen Theile der "Fauna Austriaca," mit einer näheren Begründung der in derselben aufgenommenen neuen Dipteren-Gattungen

Schiner IR (1868) Diptera. In: von Wullerstorf-Urbair B (ed) Reise der österreichischen Fregatte Novara um die Erde in den Jahren 1857, 1858, 1859 unter den Befehlen des Commodore B. von Wüllerstorf-Urbair. Zoologischer Theil, vol 2. Abt. 1. B. K. Gerald's Sohn, Wien, pp 1–388. [vi + 388 pp., 4 pls.]

Scopoli JA (1763) Entomologia carniolica exhibens insecta carnioliae indigena et distribute in ordines, genera, species, varietates. Methodo Linnaeana. Trattner, Vindobonae [=Vienna], 420 pp. https://doi.org/10.5962/bhl.title.119976

Sengupta J, Naskar A, Maity A, Hazra S, Mukhopadhyay E, Banerjee D, Ghosh S (2016a) An updated distributional account of Indian hover flies (Insecta: Diptera: Syrphidae). J Entomol Zool Stud 4:381–396

Sengupta J, Naskar A, Maity A, Hazra S, Banerjee D (2016b) New distributional records and annotated keys of hover flies (Insecta: Diptera: Syrphidae) from Himachal Pradesh India. J Adv Zool 37(1):31–54

Sengupta J, Naskar A, Maity A, Hazra S, Sarkar NK, Banerjee D (2017) Hover flies (Diptera: Syrphidae) from Darjeeling Himalaya—a part of Indo-Burmese Hotspot. Indian J Entomol 79(3):336–353

Sengupta J, Naskar A, Maity A, Banerjee D (2018a) Effects of selected environmental variable upon the distribution of hover fly (Insecta: Diptera: Syrphidae) along an altitudinal gradient. Int J Adv Life Sci Res 1(3):1–6

Sengupta J, Naskar A, Maity A, Homechaudhuri S, Banerjee D (2018b) A taxonomic account of hover flies (Insecta: Diptera: Syrphidae) with 4 new records from cold dry zones of Himachal Pradesh, India. Int J Adv Life Sci Res 1(4):13–30

Sengupta J, Naskar A, Banerjee D (2019a) Pollinating Diptera: the forgotten superhero. ENVIS Newsl 1(3):1–6

Sengupta J, Naskar A, Maity A, Homechaudhuri S, Banerjee D (2019b) A short taxonomic account of dipteran flies from the Renuka Wetland and adjacent Sanctuary, Himachal Pradesh, India. Munis Entomol Zool J 14(1):224–231
Sengupta J, Naskar A, Homechaudhuri S, Banerjee D (2019c) Diversity of hover flies (Insecta: Diptera: Syrphidae) with 3 new records from Shivalik Hill Zone of Himachal Pradesh, India. Int J Adv Life Sci Res 2(3):39–35
Sengupta J, Naskar A, Homechaudhuri S, Banerjee D (2019d) Diversity of hover flies (Insecta: Diptera: Syrphidae) with 3 new records from High Hill Zone of Himachal Pradesh, India. Int J Adv Life Sci Res 2(4):21–37
Sengupta J, Naskar A, Maity A, Parui P, Homchaudhuri S, Banerjee D (2020) A new record of the hoverfy genus Dasysyrphus Enderlein, 1938 (Insecta: Diptera: Syrphidae) from India. J Threat Taxa 12(4):15503–15506
Sengupta J, Mukherjee A, Naskar A, Banerjee D (2023a) An insight into the taxonomy and diversity of pollinating hoverflies (Insecta: Diptera: Syrphidae) from dry deciduous landscape of West Bengal. Rec Zool Surv India 123(2S):467–488
Sengupta J, Naskar A, Banerjee D (2023b) Morpho taxonomic studies of Syrphid pollinators (Insecta: Diptera) from lesser Himalayas: occasional paper no., 1–232. Published by the Director. Zoological Survey of India, Kolkata
Sengupta J, Roy B, Naskar A, Banerjee D (2024) A Catalogue of Indian Hoverflies (Insecta: Diptera: Syrphidae), Records of the Zoological Survey of India, Occasional Paper No. 420:1–198.
Senior-White R (1922) Notes on Indian Diptera. I. Notes on the Diptera of the Khasia Hills
Shah GM, Jan U, Wachkoo AA (2014) A checklist of Hoverflies (Diptera: Syrphidae) in the Western Himalaya, India. Acta Zool Acad Sci Hung 60(4):283–305
Shannon RC (1922) A reclassification of the subfamilies and genera of North American Syrphidae (Diptera)[concl.]. Bullet Brooklyn Entomol Soc 17:30–42
Shiraki T (1949) A new genus of Syrphidae, *Matsumyia*. Insecta Matsumurana 17:1–3
Shiraki T (1930) Die syrphiden des japanischen Kaiserreichs, mit Berucksichtung benachbarter gebiete. Mem Fac Sci Agric Taihoku Imp Univ 1:1–446
Singh A, Sodhi NS, Gupta V (1985) A report on the syrphid fauna of Chandigarh and surrounding areas. Res Bull Panjab Univ Sci 36:191–219
Singh P, Thakur M, Sharma KC, Sharma HK, Nayak RK (2020) Larval feeding capacity and pollination efficiency of the aphidophagous syrphids, Eupeodes frequens (Matsmura) and Episyrphus balteatus (De Geer) (Diptera: Syrphidae) on the cabbage aphid (Brevicoryne brassicae L.) (Homoptera: Aphididae) on mustard crop. Egypt J Biol Pest Control 30:105
Skevington JH, Locke MM, Young AD, Moran K, Crins WJ, Marshall SA (2019) Field guide to the flower flies of northeastern North America. Princeton University Press, Princeton
Smirnov ES (1924) Eine neue Syrphiden-Gattung aus Turkestan. Entomologische Mitteilungen 13:94–95
Sodhi NS, Singhi A (1991) Three new species of family Syrphidae (Diptera) from India. Acta Zool Cracov 34:315–322
Stackelberg AA (1952) Novye Syrphidae (Diptera) palearkticheskoy fauny. Trudy Zoologicheskogo Instituta Leningrad 12:350–400
Stackelberg AA (1974) New species of hover-flies (Diptera, Syrphidae) of Siberia and Mongolian People's Republic. Entomologicheskoe Obozrenie 53:443–446
Steenis JV, Hippa H, Mutin VA (2018) Revision of the Oriental species of the genus *Sphegina* Meigen, 1822 (Diptera: Syrphidae). Eur J Taxon 489:1–198
Stuckenberg BR (1954b) The *Paragus serratus* complex, with descriptions of new species (Diptera: Syrphidae). Trans R Entomol Soc London 105:393–422
Thomson CG (1869) Diptera Sueciae, vol 7. Lund University, Lund
Thompson CG (1869) Diptera Sueciae, vol 7. Lund University, Lund
Thompson FC (1974) The genus *Spheginobaccha* de Meijere (Diptèrea: Syrphidae). Trans Am Entomol Soc 100:255–287
Thompson FC (2000) A new Oriental *Blera* (Diptera: Syrphidae). Entomol News 111:181–184
Thompson FC, Ghorpadé K (1992) A new coffee aphid predator, with notes on other Oriental species of Paragus (Diptera: Syrphidae). Colemania 5:1–24

Thompson FC, Rotheray G (1998) Family Syrphidae. In: Papp L, Darvas B (eds) Contributions to a manual of palaearctic diptera, vol 3. Science Herald, Budapest, pp 81–139

Thompson FC, Vockeroth JR, Speight MCD (1982) The Linnaean species of flower flies (Diptera: Syrphidae). Mem Entomol Soc Wash 10:150–165

UNEP (2010) Global honey bee colony disorders and other threats to insect pollinators, UNEP emerging issues report, United Nations Environmental Program, Nairobi

Van der Sluijs JP, Vaage NS (2016) Pollinators and global food security: the need for holistic global stewardship. Food Ethics 1(1):75–91

Verrall GH (1901) Platypezidae, Pipunculidae and Syrphidae of Great Britain. British Flies 8

Violovitsh NA (1976) Survey of species of the genus *Baccha* Fabricius, 1805 (Diptera, Syrphidae) from the Palaearctic fauna. Novosti Fauny Sibiri 10:130–154

Vockeroth JR (1969) A revision of the genera of the Syrphini (Diptera: Syrphidae). Memoirs Entomol Soc Canada 101(62):1–176

Vujić A, Ståhls G, Rojo S, RADENKOVIĆ S, ŠIMIĆ S (2008) Systematics and phylogeny of the tribe Paragini (Diptera: Syrphidae) based on molecular and morphological characters. Zool J Linn Soc 152(3):507–536

Walker F (1849) List of the specimens of dipterous insects in the collection of the British Museum. Part III. British Museum, London, 1 + 203 pp. [pp. (3) + 485–687]

Walker F (1849–1856) Insecta Britannica, Diptera, vol 1–3. Lovell Reeve, London

Walker F (1852) Diptera. Part III. In: Saunders WW (ed) Insecta Saundersiana: or characters of undescribed insects in the collection of William Wilson Sauders, Esq., F.R.S., F.L.S., & c. Vol. 1. Van Voorst, London, pp 157, 474–252

Walker F (1856) Catalogue of the dipterous insects collected at Sarawak, Borneo, by Mr. A.R. Wallace, with descriptions of new species. J Proc Linn Soc 1(3):105–136. https://doi.org/10.1111/j.1096-3642.1856.tb00964.x

Walker F (1859–1860) Catalogue of the dipterous insects collected at Makessar in Celebes, by Mr. A. R. Wallace, with descriptions of new species. J Proc Linn Soc London, Zool 4:97–144.

Walker F (1861) Catalogue of the dipterous insects collected in Batchian, Kaisaa and Makian, and at Tidon in Celebes, by Mr. A. R. Wallace, with descriptions of new species. J Proc Linnaean Soc London Zool 5:270–303

Walker F (1865) Descriptions of new species of the dipterous insects of New Guinea. J Proc Linnaean Soc London Zool 8(102–108):109–130

Warren SD, Harper KT, Booth GM (1988) Elevational distribution of insect pollinators. Am Midl Nat 120(2):325–330

Westermann BW (1821) Schreiben des Herrn B. W. Westermann in Copenhagen, an den Profess. Wiedemann in Kiel. Aus dem Englischen übersetzt mit einigen Anmerkungen vom Empfänger. Magazin der Entomologie 4:411–427

Wiedemann CRW (1818) Neue Insecten vom Vorgebirge der guten Hoffnung. Zoologisches Magazin 1(2):40–48

Wiedemann CRW (1819) Beschreibung neuer Zweiflugler aus Ostindien und Afrika. Zool Mag 1(3):1–39

Wiedemann CRW (1820) Munus rectoris in Academia Christiano-Albertina iterum aditurus nova dipterorum genera. Offert iconibusque illustrat. C. F. Mohr, Kiliae Holsatorum [= Kiel], 23 pp

Wiedemann CRW (1824) Munus rectoris in Academia Christiana Albertina aditurus analecta entomologica ex Museo Regio Havniensi maximw congesta profert iconibusque illustrat 60

Wiedemann CRW (1830) Aussereuropaische zweiflugelige Insekten. Zweiter Theil, Schulz, Hamm, p 684

Williston SW (1887) Synopsis of the North American Syrphidae. Bullet USA Museum 31[1886], xxx + 335:12 pls. [before 1887.06.15]

Wyatt NP (1991) Notes on Citrogramma Vockeroth (Diptera: Syrphidae) with descriptions of five new species. Orient Insects 25(1):155–169

Wyss Institute (2016) The RoboBees project and colony collapse disorder. Accessed 17 Mar 2016

Zetterstedt JW (1843) Diptera Scandinaviae. Disposita et descripta, vol 2. Officina Lundbergiana, Lundae [=Lund]. 409 pp. [pp. 441–894]

Biodiversity of Big-Headed Flies (Diptera: Pipunculidae)

3

Bristi Roy, Jayita Sengupta, Atanu Naskar, Shyamasree Ghosh, and Dhriti Banerjee

Abstract

The Pipunculidae (Insecta: Diptera), or big-headed flies, constitute a distinctive family of Diptera characterized by their disproportionately large, hemispherical head, dominated by compound eyes, and their specialized parasitic life cycle. Despite their unique morphology, ecological significance, and global occurrence, pipunculids remain poorly studied in India, with diversity assessments hindered by sparse collections and the difficulty of identifying females. This chapter reviews the diversity of Indian Pipunculidae, with an updated checklist of Indian species based on published literature. Emphasis is placed on their endoparasitism of leafhoppers (Auchenorrhyncha), potential role in biological control, and the need for integrated taxonomic, molecular, and ecological studies.

Keywords

Syrphoidea · Auchenorrhyncha · Biological control · Endoparasitoid · Integrated taxonomy

---

B. Roy · J. Sengupta (✉) · A. Naskar
Zoological Survey of India, New Alipore, Kolkata, West Bengal, India

S. Ghosh
School of Biological Sciences, National Institute of Science Education and Research (NISER), Jatni, Odisha, India

D. Banerjee
Diptera, Zoological Survey of India (ZSI), Kolkata, West Bengal, India

S. Ghosh, D. Banerjee (eds.), *Pollinators: Life, Biology and Threats*,
https://doi.org/10.1007/978-981-95-0359-9_3

Fig. 3.1 Order Diptera: Family Pipunculidae. (Photo courtesy: Dr. Supatim Laha)

3.1 Introduction

The members of the family Pipunculidae (Insecta: Diptera) are readily recognized by their globular head, which in most species is almost entirely occupied by the compound eyes, and their slender, cylindrical body form. Adults are small to medium-sized (2–12 mm) and often exhibit subdued dark coloration and cryptic habits (Claude 2025). The wing venation is distinctive, with wing cell r_5 open and the absence of a spurious vein, a key feature separating them from Syrphidae (Rafael et al. 2023). Adult pipunculids are non-flower visitors, active in sunny weather, occurring in a wide range of habitats—forests, grasslands, meadows, wetlands, etc. (Kapoor et al. 1987; Claude 2025). Unlike syrphids, they are not regular flower visitors; instead, adults can be typically encountered resting on foliage or flying low among vegetation (Kapoor et al. 1987) (Fig. 3.1).

Syrphidae and Pipunculidae were historically united as the superfamily Syrphoidea based on shared morphological features—notably similar antenna structure (three-segmented/aristate antenna), broadly similar wing venation, and comparable male head/eye morphology (males often holoptic), and perceived homologies in the male terminalia that earlier authors treated as synapomorphies. However, pipunculids differ sharply in larval biology (they are specialized endoparasitoids of Auchenorrhyncha). Eventually, subsequent phylogenetic studies place Pipunculidae as sister to Schizophora, rendering the traditional superfamily Syrphoidea paraphyletic (Guo et al. 2022).

At present, the family Pipunculidae is divided into four subfamilies—Nephrocerinae, Protonephrocerinae, Chalarinae, and Pipunculinae, encompassing about 25 extant genera and over 1484 described species worldwide (Claude 2025), of which Indian Pipunculidae is represented by 84 species in 11 genera under 2 subfamilies (Rafael and De Meyer 1992).

3.2 A Brief History

The first documentation of Indian Pipunculidae was provided by Brunetti (1912), who described 21 Oriental species, including 12 species from India. Subsequent work by Brunetti (1913, 1923) made substantial contributions to the knowledge of Indian pipunculid diversity. Following Brunetti, significant advances were made by Hardy (1949, 1956, 1966, 1972, 1975) through his studies on Oriental Pipunculidae, which included Indian representatives. Later, Kapoor et al. (1980) and Kapoor and Grewal (1985) contributed notably to the Indian faunal accounts, culminating in Kapoor et al. (1987) comprehensive monograph, the very first dedicated work on Indian pipunculids, describing over 50 new species.

A more detailed discussion on the taxonomic position of the family was presented by Rafael and De Meyer (1992). Subsequent regional and systematic revisions refined generic placements for several Indian species (Skevington and Yeates 2001; Skevington 2002, 2005; Kehlmaier 2005). Recent multigene molecular phylogenies have further reshaped higher-level relationships and emphasized that Indian taxa remain underrepresented in molecular datasets (Skevington and Yeates 2000; Pauli et al. 2018; Guo et al. 2022).

3.3 Parasitism and Host Range

Pipunculid larvae are endoparasitoids of Auchenorrhyncha (Hemiptera), especially Cicadellidae, Cixiidae, Cercopidae, Flatidae, Fulgoridae, Delphacidae, and Membracidae, with the notable exception of the genus *Nephrocerus* Zetterstedt, 1838, whose members parasitize adult crane flies (Tipulidae) (Claude 2025). Females insert eggs into nymph or adult hosts using their piercing ovipositors. The larva develops within the host, eventually killing it and emerging to pupate externally.

This parasitic strategy indicates a potential role in the biological control of leafhoppers in agriculture. However, host associations are documented for only a few Indian species to date (Brunetti 1923; Kapoor et al. 1987).

3.4 Collection and Preservation of Big-Headed Flies

Random sweep-netting over low vegetation during sunny periods is often the most effective method for collecting pipunculids, as they are infrequent and not easily detected. Passive collection methods, such as malaise traps and yellow pan traps, can also be employed. Captured specimens need to be narcotized using ethyl acetate soaked in cotton within a killing jar. For further studies and preservation, specimens may be stored either in vials containing 70% ethanol for molecular analyses or in paper envelopes with camphor crystals and a few drops of camphor–phenol mixture (ratio: 3:1) for morphological studies, ensuring all specimens are accompanied by complete collection data.

3.5 Systematic List of Indian Pipunculid Fauna (After Rafael & De Meyer 1992)

Family PIPUNCULIDAE
Subfamily CHALARINAE
Genus *Verrallia* Mik, 1899

1. *Verrallia plumbella* (Brunetti, 1912)

Subfamily PIPUNCULINAE
Genus *Cephalops* Fallén, 1810

2. *Cephalops flavocinctus* (Brunetti, 1912)
3. *Cephalops kashmerensis* Kapoor, Grewal & Sharma, 1987
4. *Cephalops kumaonensis* Kapoor, Grewal & Sharma, 1987
5. *Cephalops mashobraensis* Kapoor, Grewal & Sharma, 1987
6. *Cephalops nigronitens* (Brunetti, 1912)

Subgenus *Semicephalops* De Meyer, 1994

7. *Cephalops (Semicephalops) chandiensis* (Kapoor, Grewal & Sharma, 1987)
8. *Cephalops (Semicephalops) eufraternus* (Kapoor, Grewal & Sharma, 1987)

Genus *Cephalosphaera* Enderlein, 1936

9. *Cephalosphaera prolata* (Kertész, 1915)
10. *Cephalosphaera motichoorensis* Kapoor, Grewal & Sharma, 1987

Genus *Claraeola* Aczél, 1940

11. *Claraeola suranganiensis* (Kapoor & Grewal, 1985)

Genus *Clistoabdominalis* Skevington and Yeates, 2001

12. *Clistoabdominalis macropygus* (Meijere, 1914)
13. *Clistoabdominalis nitidifrons* (Becker, 1900)
14. *Clistoabdominalis roralis* (Kertész, 1915)
15. *Clistoabdominalis uniformis* (Brunetti, 1917)

Genus *Dasydorylas* Skevington and Yeates, 2001

16. *Dasydorylas orientalis* (Koizumi, 1959)

Genus *Dorylomorpha* Aczél, 1939

Subgenus *Dorylomyia* Albrecht, 1990

17. *Dorylomorpha (Dorylomyia) abdochaetus* Kapoor, Grewal & Sharma, 1987
18. *Dorylomorpha (Dorylomyia) appendiculata* Kapoor, Grewal & Sharma, 1987

Subgenus *Dorylomima* Albrecht, 1990

19. *Dorylomorpha (Dorylomima) abdoflavus* Kapoor, Grewal & Sharma, 1987
20. *Dorylomorpha (Dorylomima) indica* Albrecht, 1990

Genus *Eudorylas* Aczél, 1940

21. *Eudorylas acroacanthus* (Hardy, 1968)
22. *Eudorylas albucus* (Hardy, 1968)
23. *Eudorylas antennalis* Kapoor, Grewal & Sharma, 1987
24. *Eudorylas bharatiensis* Kapoor, Grewal & Sharma, 1987
25. *Eudorylas bicostalis* Kapoor, Grewal & Sharma, 1987
26. *Eudorylas biroi* (Kertész, 1903)
27. *Eudorylas corbetti* Kapoor, Grewal & Sharma, 1987
28. *Eudorylas dextrostylus* Kapoor, Grewal & Sharma, 1987
29. *Eudorylas euciliatus* Kapoor, Grewal & Sharma, 1987
30. *Eudorylas eulentiger* Kapoor, Grewal & Sharma, 1987
31. *Eudorylas extensus* (Brunetti, 1923)
32. *Eudorylas filicornis* (Brunetti, 1912)
33. *Eudorylas garhwalensis* Kapoor, Grewal & Sharma, 1987
34. *Eudorylas indiaensis* Kapoor, Grewal & Sharma, 1987
35. *Eudorylas jacksoni* Kapoor, Grewal & Sharma, 1987
36. *Eudorylas javanensis* (Brunetti, 1923)
37. *Eudorylas khaziarensis* Kapoor, Grewal & Sharma, 1987
38. *Eudorylas ludhianaensis* Kapoor, Grewal & Sharma, 1987
39. *Eudorylas luteipes* (Brunetti, 1923)
40. *Eudorylas major* (Brunetti, 1923)
41. *Eudorylas manaliensis* Kapoor, Grewal & Sharma, 1987
42. *Eudorylas minghlanii* Kapoor, Grewal & Sharma, 1987
43. *Eudorylas mutillatus* Loew, 1858
44. *Eudorylas pennatus* Kapoor, Grewal & Sharma, 1987
45. *Eudorylas pinjorensis* Kapoor, Grewal & Sharma, 1987
46. *Eudorylas poantaensis* Kapoor, Grewal & Sharma, 1987
47. *Eudorylas quartarius* (Brunetti, 1912)
48. *Eudorylas scutellatus* Kapoor, Grewal & Sharma, 1987
49. *Eudorylas thekkadiensis* Kapoor, Grewal & Sharma, 1987
50. *Eudorylas tibiatus* Kapoor, Grewal & Sharma, 1987
51. *Eudorylas tuberculatus* Kapoor, Grewal & Sharma, 1987

Genus *Microcephalops* De Meyer, 1990

52. *Microcephalops subaeneus* (Brunetti, 1923)
53. *Microcephalops subdolosus* (Kapoor, Grewal & Sharma, 1987)
54. *Microcephalops kurseongiensis* (Kapoor, Grewal & Sharma, 1987)

Genus *Pipunculus* Latreille, 1802

55. *Pipunculus himalayensis* Brunetti, 1912
56. *Pipunculus annulifemur* Brunetti, 1923
57. *Pipunculus mirabilis* Brunetti, 1912

Genus *Tomosvaryella* Aczél, 1939

58. *Tomosvaryella baderiensis* Kapoor, Grewal & Sharma, 1987
59. *Tomosvaryella curvipatis* Kapoor, Grewal & Sharma, 1987
60. *Tomosvaryella dehraduniensis* Kapoor, Grewal & Sharma, 1987
61. *Tomosvaryella denticulatus* Kapoor, Grewal & Sharma, 1987
62. *Tomosvaryella epichalca* (Perkins, 1905)
63. *Tomosvaryella equistylis* Kapoor, Grewal & Sharma, 1987
64. *Tomosvaryella eusylvatica* Kapoor, Grewal & Sharma, 1987
65. *Tomosvaryella genitalis* Kapoor, Grewal & Sharma, 1987
66. *Tomosvaryella guwahatiensis* Kapoor, Grewal & Sharma, 1987
67. *Tomosvaryella indica* Kapoor, Grewal & Sharma, 1987
68. *Tomosvaryella kashipurensis* Kapoor, Grewal & Sharma, 1987
69. *Tomosvaryella limpidipennis* (Brunetti, 1912)
70. *Tomosvaryella longipygiamis* Kapoor, Grewal & Sharma, 1987
71. *Tomosvaryella longistyla* Kapoor, Grewal & Sharma, 1987
72. *Tomosvaryella minutus* Kapoor, Grewal & Sharma, 1987
73. *Tomosvaryella multisetae* Kapoor, Grewal & Sharma, 1987
74. *Tomosvaryella nitens* (Brunetti, 1912)
75. *Tomosvaryella oryzaetora* Koizumi, 1959
76. *Tomosvaryella pterae* (Kapoor & Grewal, 1985)
77. *Tomosvaryella ramnagariensis* Kapoor, Grewal & Sharma, 1987
78. *Tomosvaryella robusta* Hardy, 1968
79. *Tomosvaryella subhectena* Kapoor, Grewal & Sharma, 1987
80. *Tomosvaryella subrobusta* Kapoor, Grewal & Sharma, 1987
81. *Tomosvaryella surstylae* Kapoor, Grewal & Sharma, 1987
82. *Tomosvaryella sylvatica* (Meigen, 1824)
83. *Tomosvaryella tattapaniensis* Kapoor, Grewal & Sharma, 1987
84. *Tomosvaryella troangulatus* Kapoor, Grewal & Sharma, 1987

3.6 Discussion

Worldwide, over 1484 pipunculid species are reported (Claude 2025), of which India houses only 84 species in 11 genera under 2 subfamilies (Rafael and De Meyer 1992). The present synthesis underscores that, despite over a century of recorded study, the Indian Pipunculidae remain severely underrepresented in both faunistic and ecological datasets. Historical accounts—from Brunetti's foundational descriptions in the early twentieth century through Kapoor's landmark 1987 monograph—have established a baseline of species diversity; yet, Indian pipunculid research has not been substantially expanded in the recent few decades. The paucity of new records reflects both the cryptic nature of adults and the absence of targeted sampling strategies, particularly in biodiversity-rich but poorly explored regions such as the Northeastern Himalayas, Western Ghats, and central Indian forested landscapes.

Pipunculidae, with their specialized parasitism of Auchenorrhyncha, have strong potential in agricultural pest management, especially against leafhoppers and planthoppers that transmit plant pathogens. Yet, host association data for Indian taxa are scarce—Kapoor et al. (1987) enlisted 81 known pipunculids and their hosts worldwide, which included only three Indian species, namely, *Pipunculus annulifemur* Brunetti, 1923; *Tomosvaryella sylvatica* (Meigen, 1824); and *Tomosvaryella troangulatus* Kapoor, Grewal & Sharma, 1987, all parasitize Cicadellidae. The lack of comprehensive host records, standardized sampling methods, and major surveys since Kapoor et al. (1987) hinders the assessment of the ecological and biocontrol value of the Indian pipunculids.

Expanding taxon sampling from India will not only fill significant geographic gaps in molecular datasets but also contribute to a more accurate reconstruction of pipunculid evolutionary history and biogeography. Therefore, pipunculids should be prioritized in future entomological investigations.

References

Aczél ML (1939) Das System der Familie Dorylaidae. Dorylaiden-Studien I. Zool Anz 125:15–23

Aczél ML (1940) Vorarbeiten zu einer Monographie der Dorylaiden (Diptera). Dorylaiden-Studien V. Zool Anz 132:149–169

Albrecht A (1990) Revision, phylogeny and classification of the genus *Dorylomorpha* (Diptera, Pipunculidae). Acta Zool Fenn 188:1–240

Becker T (1900) Dipterologische Studien V. Pipunculidae. Erste Fortsetzung. Berliner Entomologische Zeitschrift 45:215–252

Brunetti E (1912) Diptera Nematocera (excluding Chironomidæ and Culicidæ), Vol. 1. London: Taylor & Francis

Brunetti E (1913) Zoological results of the Arbor Expedition, 1911–12. XI. Diptera. Rec Indian Museum 8:149–190

Brunetti E (1917) Diptera of the Simla district. Rec Indian Museum 13:59–102

Brunetti E (1923) Diptera Brachycera, Vol. 3. London: Taylor & Francis

Claude J (2025) About the Pipunculidae (Diptera) at the Musée des confluences in Lyon (France). Dipterists digest 32:58–68

de Meijere JCH (1914) Studien über sudostasiatische Dipteren. IX. Tijdschrift voor Entomologie 57:137–168

De Meyer M (1990) Systematics of the Nearctic species of the genus *Cephalops* Fallen (Diptera, Pipunculidae). Bull Inst R Sci Nat Belg Ent (1989) 69:99–130

De Meyer M (1994) Phylogenetic relationships within the Cephalopsini (Diptera, Pipunculidae). Bulletin et Annales de la Société Entomologique de Belgique 130:7–18

Enderlein G (1936) 22. Ordnung Zweiflugler, Diptera. In: Brohmer P, Ehrmann P, Ulmer G (eds) Die Tierwelt Mitteleuropas. 6 (2), Insekten, Teil 3, pp. 259

Fallen CF (1810) Specim. entomolog. novam Diptera disponendi methodum exhibens. Berlingianis, Lund, p 26

Guo S, Chen J, Song N (2022) Phylogenomic analysis of Syrphoidea (Diptera: Syrphidae, Pipunculidae) based on the expanded mitogenomic data. Arch Insect Biochem Physiol e21998. https://doi.org/10.1002/arch.21998

Hardy DE (1949) New Dorilaidae from the Belgian Congo. Bulletin del'Institut Royal des Sciences Naturelles de Belgique, Entomologie 25(39):1–10

Hardy DE (1956) Diptera: Dorilaidae (Pipunculidae). Insects of Micronesia 13(1):1–9

Hardy DE (1966) Diptera from Nepal. Pipunculidae (Dorilaidae). Bullet Br Museum (Nat Hist) Entomol 17(10):439–449

Hardy DE (1968) Bibionidae and Pipunculidae of the Philippines and Bismarck Islands (Diptera). Entomologiske Meddelelser 36:417–507

Hardy DE (1972) Studies on Oriental Pipunculidae (Diptera). Orient Insect Suppl nr-2:1–76

Hardy DE (1975) Family Pipunculidae. In: Delfinado MD, Hardy DE (eds) A catalogue of Diptera of the Oriental Region II. University Press of Hawaii, Honolulu, pp 296–306

Kapoor VC, Grewal JS (1985) On a new subgenus of *Pipunculus* and a new species of *Dorylomorpha* (Diptera: Pipunculidae) from India. Int J Entomol 27:398–403

Kapoor VC, Agarwal ML, Grewal JS (1980) Two new species of Pipunculidae (Diptera: Pipunculidae) from Raniket, Kumaon hills, India. Bullet Entomol 18(1977):74–77

Kapoor VC, Grewal JS, Sharma SK (1987) Indian pipunculids (Diptera: Pipunculidae). A comprehensive monograph. Atlantic Publishers & Distributors, New Delhi, p 201

Kehlmaier C (2005) Taxonomic studies on Palaearctic and Oriental Eudorylini (Diptera: Pipunculidae), with the description of three new species. Zootaxa 1030:1–48

Kertész K (1903) Die *Pipunculus*-Arten Sud Asiens und Neu-Guineas. Annales Historico-Naturales Musei Nationalis Hungarici 1:465–471

Kertész K (1915) Contributions to the knowledge of thc Dorylaidae. Annales Historico-Naturales Musei Nationalis Hungarici 13:386–392

Koizumi K (1959) On four dorilaid parasites of the green rice leafhopper, *Nephotettixcincticeps* Uhler (Diptera). Sci Rep Facul Agric Okayama Univ 13:37–45

Latreille PA (1802) Histoire naturelle, generale et particuliere, des crustaces et des insectes. Tome troisieme. Dufart, Paris. xii + 13-467 + 1 pp

Loew H (1858) Bidrag till kannedomen om Afrikas Diptera [part]. Öfvers K Vet Akad Förh 14:337–383

Meigen JW (1824) Systematische Beschreibung der bekannten europäischen zweiflügeligen Insekten. Vierter Theil. Schulz-Wundermann, Hamm. xii + 428 pp

Mik J (1899) *Verrallia* nov. gen. Pipunculidarum (Dipt.). Wiener Entomologische Zeitung 18:133–137

Pauli T, Burt TO, Meusemann K, Bayless K, Donath A, Podsiadlowski L, Mayer C, Kozlov A, Vasilikopoulos A, Liu S, Zhou X, Yeates D, Misof B, Peters RS, Mengual X (2018) New data, same story: phylogenomics does not support Syrphoidea (Diptera: Syrphidae, Pipunculidae): Syrphoidea are not supported by new data. Syst Entomol 43(3):447–459

Perkins RCL (1905) Leafhoppers and their natural enemies. (Pt. IV. Pipunculidae). Bulletin Division of Entomology Hawaiian Sugar Planters Association Experiment Station 1:119–157

Rafael JA, De Meyer M (1992) Generic classification of the family Pipunculidae (Diptera): a cladistic analysis. J Nat Hist 26:637–658

Rafael JA, González CR, Ale-Rocha R (2023) A catalog of Pipunculidae of Chile (Diptera). Papéis Avulsos de Zoologia 63:e202363017

Skevington JH (2002) Phylogenetic revision of Australian members of the *Allomethus* genus group (Diptera: Pipunculidae). Insect Syst Evol 33:133–161

Skevington JH (2005) Revision of Nearctic *Nephrocerus* Zetterstedt (Diptera: Pipunculidae). Zootaxa 977:1–36

Skevington JH, Yeates DK (2000) Phylogeny of the Syrphoidea (Diptera) inferred from mtDNA sequences and morphology with particular reference to classification of the Pipunculidae (Diptera). Mol Phylogenet Evol 16(2):212–224

Skevington JH, Yeates DK (2001) Phylogenetic classification of Eudorylini (Diptera: Pipunculidae). Syst Entomol 26:421–452

Zetterstedt JW (1838) Sectio tertia. Diptera. Dipterologis Scandinaviae, pp. 477–868. In his: Insecta Lapponica. Lipsiae [= Leipzig]. vi + 1,140 pp. [before1838.09.01]

Plant-Pollinator Interaction

4

Shyamasree Ghosh and Md. Nurul Hasan

Abstract

Pollinators are vital components of terrestrial ecosystems, underpinning plant reproduction and maintaining biodiversity through their pollination and ecosystem service activity. This ecological service supports over 87% of flowering plant species and is essential for the productivity of more than 75% of global crop species. The intricate mutualisms between plants and their pollinators have evolved over millions of years, resulting in a complex web of co-adapted traits that ensure reproductive success and ecological resilience. However, pollinator populations worldwide are facing unprecedented threats from habitat destruction, pesticide use, climate change, disease, and invasive species. These pressures are not only disrupting plant-pollinator interactions but also endangering food security and ecosystem stability. The ecological and economic importance of pollinators necessitates urgent conservation measures. In this chapter we discuss (i) the biology and diversity of pollinators, (ii) the mechanisms of floral adaptation, (iii) the ecological and agricultural significance of pollination and (iv) outline the multifaceted threats faced by pollinators and present integrated conservation strategies including habitat restoration, policy interventions, scientific monitoring, and public engagement. Through a synthesis of current research and case studies, this work emphasizes the need for collaborative, science-based approaches in protecting pollinators and their health and thereby protecting the essential services provided by pollinators to both natural ecosystems and human societies.

S. Ghosh (✉)
School of Biological Sciences, National Institute of Science Education and Research (NISER), Jatni, Odisha, India

M. N. Hasan
Zoological Survey of India (Ministry of Environment, Forest and Climate Change, Government of India), Kolkata, West Bengal, India

S. Ghosh, D. Banerjee (eds.), *Pollinators: Life, Biology and Threats*,
https://doi.org/10.1007/978-981-95-0359-9_4

Keywords

Pollinators · Pollination · Biodiversity · Conservation · Ecosystem services · Threats

4.1 Introduction

Pollination represents one of the most critical biological interactions in terrestrial ecosystems, leading to the reproductive success of angiosperms and supporting the ecological networks in which they exist. It involves the transfer of pollen grains from the male anthers of a flower to the female stigma, leading to fertilization and the production of seeds. While abiotic factors such as wind and water contribute to pollination in some plant species, biotic pollination, primarily mediated by animals, especially insects, is by far the most prevalent and evolutionarily significant mechanism.

Animal-mediated pollination is estimated to be essential for the reproduction of approximately 87.5% of flowering plant species globally (Ollerton et al. 2011), highlighting the indispensable role pollinators play in maintaining biodiversity.

The plant-insect pollinator interaction (Figs. 4.1, 4.2, and 4.3) is a vital ecological relationship that plays a key role in the reproduction of many plant species. It encompasses the following:

Flowers and the attraction of insects	Plants evolved with colorful, fragrant flowers, offering nectar and pollen, attracting pollinators, like bees, butterflies, beetles, moths, or flies
A pollinator visits the plant	Insects are attracted to the flowers for food, like nectar or pollen. On landing on a flower to feed, their bodies brush against the flower's anthers, the male reproductive part, leading to the picking up of pollen
Transfer of pollen	Insects moving from flower to flower, cross-pollinate, transfer the pollen from the anther of one flower to the stigma, the female reproductive part of another flower, ensuring genetic diversity of plants, and are critical for the production of fruits and seeds in many species
Benefits to the pollinator	Insects benefit from interaction by obtaining nectar for energy or pollen for food for their larvae, fulfilling their role in the plant's reproductive process
Plant reproduction	Cross-pollination increases fertilization likelihood, leading to seed production, essential for the plant's reproduction and spread

The relationship between plants and insect pollinators is a form of mutualism, as beneficial both for the plant and the pollinators. This mutualistic relationship reveals co-evolution over more than 100 million years, leading to the formation of a complex web of ecological and morphological specializations in both plants and their pollinators (van der Niet et al. 2011). Plants have evolved a wide range of floral traits of color, fragrance, nectar composition, and flower morphology to attract specific pollinators, while pollinators have concurrently developed specialized structures, sensory mechanisms, and behaviors to efficiently exploit floral rewards (Thomson 2003).

Fig. 4.1 Varying symmetry and relative benefits for flower visitors and plants. The horizontal bar depicts the relative benefit for the pollinator (left) and the plant (right). (Image adapted with permission from van der Kooi et al. (2021))

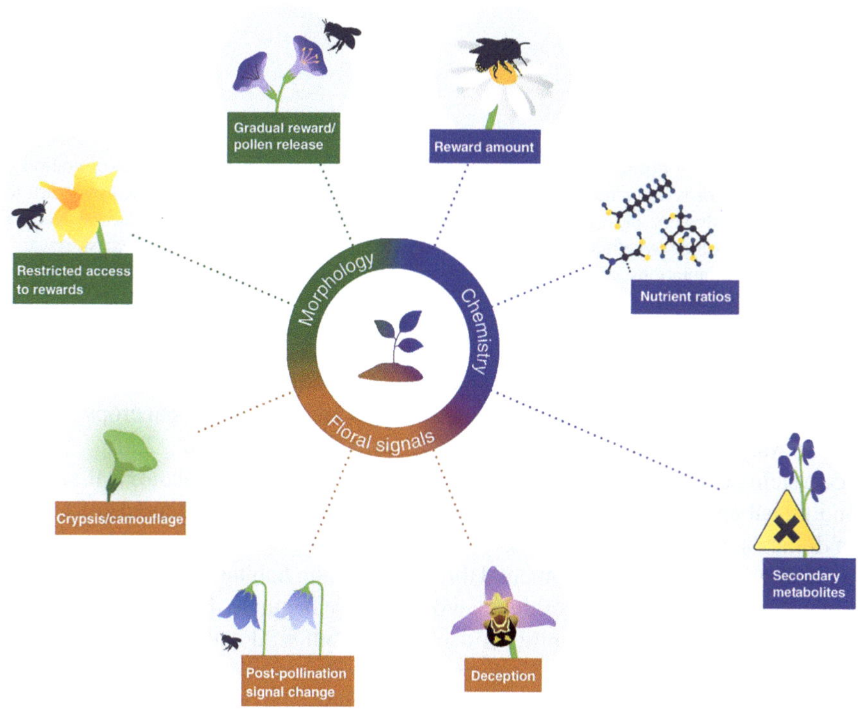

Fig. 4.2 Exemplary plant adaptations to increase advantage from interaction with pollinators. Characteristics of the reward chemistry can render the reward quantity or quality suboptimal, and thereby deter certain flower visitors or incentivize a return visit. Signals acting on pollinator senses can increase or decrease floral attraction to pollinators and other flower visitors. Flower morphology and biomechanics determine the handling time and accessibility of rewards (van der Kooi et al. 2021)

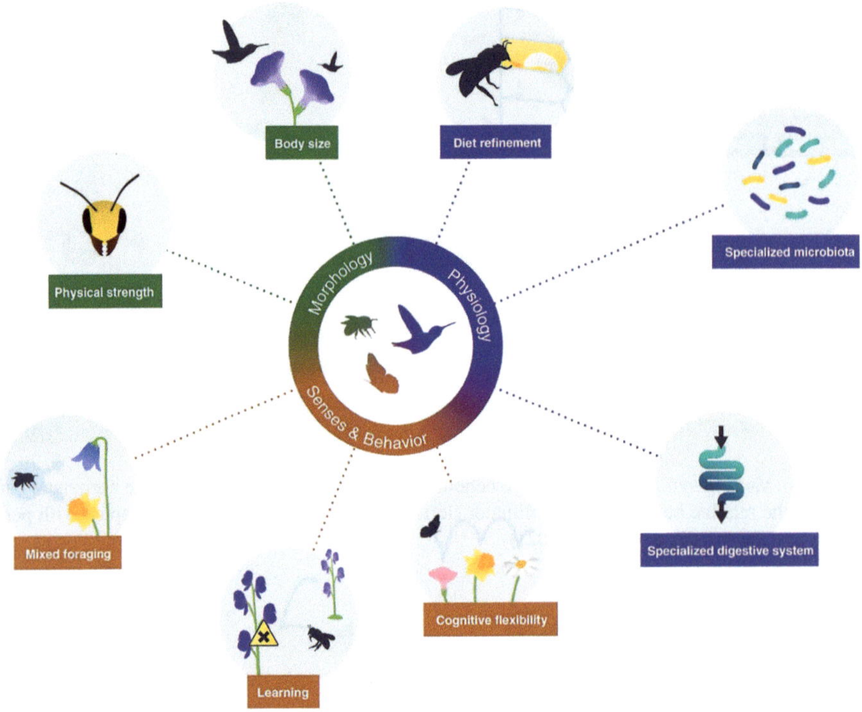

Fig. 4.3 Exemplary pollinator adaptations and behavioral responses to increase exploitation of flowers. Pollinator senses and foraging behavior determine how easily a pollinator can (learn to) detect or avoid flowers and associate signals with a reward. Morphological adaptations, such as body size, determine whether a pollinator can reach and enter the flower and access rewards. Physiological adaptations allow pollinators to efficiently exploit the rewards and to cope with reward toxicity

These interactions are not only fundamental to natural ecosystems but also critically support human food production, with over 75% of major global crops depending to varying degrees on pollination by animals (Klein et al. 2007a, b). The ecosystem service provided by pollinators thus supports both ecological stability and global agricultural economies, with an estimated value of hundreds of billions of dollars annually (IPBES 2016).

However, growing environmental threats, including habitat destruction, pesticide use, climate change, and emerging pathogens, have begun to destabilize this essential relationship, causing significant declines in both wild and managed pollinator populations (Potts et al. 2010; Goulson et al. 2015).

Therefore, understanding the biology, ecology, and vulnerabilities of plant-pollinator interactions is a crucial step toward developing sustainable strategies for biodiversity conservation and food security in the face of global environmental change.

4.2 Types of Pollinators

Pollinators encompass a diverse array of animals that transfer pollen as they forage for floral rewards such as nectar or pollen, inadvertently aiding in plant reproduction. The major pollinator groups include insects such as bees, butterflies, moths, flies, and beetles, birds including hummingbirds and sunbirds, bats, and several other small mammals.

Among these, insects, especially bees under Order: Hymenoptera, are the most significant and efficient pollinators in both natural and agricultural ecosystems due to their activity of floral visitation, specialized body structures for pollen collection, and strong floral fidelity (Klein et al. 2007a, b).

Bees possess branched body hairs and corbiculae or pollen baskets that enable efficient pollen transfer, and their complex social behavior in species like honey bees (Apis spp.) contributes to high pollination rates across multiple plant species.

Butterflies and moths under the Order Lepidoptera are also important pollinators, particularly for flowers with long, tubular corollas that are often inaccessible to bees. These insects are visually oriented, primarily attracted to brightly colored flowers, and are vital pollinators for various diurnal and nocturnal plant species. Moths, especially hawkmoths, play a significant role in pollinating nocturnally blooming plants, many of which emit strong fragrances to attract these night-time visitors (Fenster et al. 2004).

Birds, notably hummingbirds of America and sunbirds in Africa and Asia, are key pollinators of ornithophilous plants. These birds are typically attracted to bright red or orange flowers that are rich in nectar but often lack fragrance. Their long, slender bills and hovering abilities make them well-suited to access nectar from deep floral tubes, and in doing so, they facilitate cross-pollination (Cronk and Ojeda 2008).

Bats, though less commonly thought of in this context, are crucial pollinators in tropical and arid regions. They pollinate a variety of nocturnal and large-flowered plant species, including many economically important plants like agave and durian. Bat-pollinated flowers tend to be pale or white, open at night, and emit strong fruity or musky odors to attract chiropteran pollinators (Fleming et al. 2009).

In addition to these primary groups, other animals such as beetles, flies, wasps, lemurs, and even reptiles contribute to pollination in specialized ecosystems. Each pollinator group interacts with plants through unique morphological and behavioral traits, forming intricate ecological networks that enhance plant diversity and resilience.

4.3 Mechanisms of Plant Adaptation

Plants have developed a wide array of morphological, visual, and chemical traits to optimize pollination by specific animal vectors, resulting in co-evolutionary relationships that enhance reproductive success. One of the most evident adaptations involves visual cues, with flower coloration playing a vital role in attracting targeted pollinator groups.

For instance, bees are particularly attracted to blue and yellow hues, which they can perceive well due to their visual spectrum, while red-colored flowers often appeal to avian pollinators, especially hummingbirds, which possess excellent red color discrimination (Chittka and Raine 2006).

In addition to color, many flowers display nectar guides, visible or UV-reflective patterns on petals that direct pollinators toward the reproductive organs of the flower, ensuring effective pollen transfer (Leonard et al. 2011).

Another critical aspect of pollinator attraction lies in olfactory signals. Many flowers emit volatile organic compounds (VOCs) to lure pollinators, with sweet, strong, or musky fragrances being particularly effective in attracting nocturnal pollinators such as moths and bats (Raguso 2008).

These fragrance-based cues are often species-specific, helping to reduce interspecies pollen loss and ensuring that pollinators visit multiple flowers of the same species.

Plants also offer floral rewards in the form of nectar and pollen, which act as incentives for repeated visitation. Nectar, rich in sugars, provides a high-energy food source for many pollinators, while pollen offers protein, especially for bees that use it to feed their larvae. The availability and placement of these rewards are carefully aligned with floral morphology, which includes the shape, size, orientation, and symmetry of floral structures.

These morphological traits often co-evolve with the anatomical features of their pollinators, a phenomenon known as pollination syndrome, where specific floral forms correlate with certain pollinator types (Fenster et al. 2004). For example, tubular flowers favor long-tongued pollinators like butterflies and hummingbirds, while bowl-shaped flowers may be suited for short-tongued insects like beetles. Collectively, these adaptations demonstrate the sophisticated mechanisms by which plants maximize their reproductive efficiency through effective pollinator attraction and utilization.

4.4 Ecological Significance

Pollinators play a pivotal role in maintaining the structural and functional integrity of terrestrial ecosystems through their direct involvement in plant reproductive processes. By facilitating the transfer of pollen between flowers, they enable sexual reproduction in the vast majority of angiosperms, which leads to seed and fruit production, and ultimately sustains plant populations across generations. This reproductive success ensures the persistence of plant communities, which form the foundational basis of most terrestrial food webs (Ollerton et al. 2011).

Pollinators also promote gene flow within and among fragmented plant populations, thereby preserving genetic diversity, reducing inbreeding, and enhancing the adaptive capacity of species in changing environments (Memmott et al. 2004).

Genetic diversity within plant communities is directly linked to ecosystem resilience, as it increases a habitat's ability to withstand environmental stressors like climate change, invasive species, and habitat disturbance. Beyond reproductive services, pollinators contribute indirectly to broader ecosystem functions. The success of pollinated plants supports a cascade of ecological interactions by providing food and shelter to herbivores, frugivores, and decomposers, thus maintaining trophic dynamics and habitat complexity (Kremen et al. 2007).

Pollination services also influence nutrient cycling and soil stabilization, as healthy plant communities improve litter input, microbial activity, and water regulation. Moreover, in ecosystems where keystone plant species are pollination-dependent, the presence or absence of pollinators can determine community composition and ecosystem structure (Potts et al. 2010).

The decline or loss of pollinators can therefore lead to reduced plant reproduction, diminished plant diversity, and cascading effects throughout the ecosystem, threatening the stability of ecological networks. Given their integral ecological roles, pollinators are rightly regarded as bioindicators of ecosystem health, and their conservation is critical to sustaining biodiversity and ecosystem services at local and global scales.

4.5 Agricultural Importance

Pollination is a cornerstone of global agricultural productivity, directly influencing the yield, quality, and stability of many food crops. While some crops are wind- or self-pollinated, a significant proportion, more than 75% of the world's leading crop species, depend at least partially on animal pollinators, particularly bees (Klein et al. 2007a, b).

This includes a wide variety of fruits, vegetables, nuts, and oilseeds, such as apples, almonds, blueberries, cucumbers, melons, cocoa, and coffee. For many of these crops, animal-mediated pollination is not just beneficial but essential for optimal fruit set, size, uniformity, and nutritional value (Garibaldi et al. 2013). In crops like almonds, yields are almost entirely dependent on bee activity, while in others like coffee and cocoa, pollinators significantly enhance both quantity and quality. The economic valuation of pollination services underscores their critical importance to food systems.

According to the Intergovernmental Science-Policy Platform on Biodiversity and Ecosystem Services (IPBES 2016), animal pollination contributes between US $235 and $577 billion annually to global crop production. This contribution is even more vital in developing regions, where many smallholder farmers grow pollination-dependent crops for subsistence or local markets. Beyond direct yield, pollinators also support crop diversity and food security, as many nutritionally important foods rich in vitamins, antioxidants, and other health-promoting compounds are pollination-dependent (Eilers et al. 2011).

The decline of pollinators due to habitat loss, pesticide use, climate change, and disease thus poses a serious threat not only to ecosystems but also to global agriculture and human well-being. Ensuring the conservation of pollinator diversity and abundance is, therefore, not only an ecological necessity but a strategic priority for sustainable food production and economic resilience.

4.6 Threats to Pollinators

Pollinators are increasingly vulnerable to a wide array of anthropogenic and environmental pressures that threaten their survival and effectiveness, with serious implications for ecosystem health and food security. One of the most pervasive threats is habitat loss and fragmentation, primarily driven by urbanization, intensive agriculture, and infrastructure development. These changes reduce the availability of nesting sites, diminish floral diversity, and isolate pollinator populations, undermining their reproductive success and long-term viability (Winfree et al. 2009).

As natural landscapes are converted into monocultures or urban sprawl, pollinators experience declines in both food resources and habitat connectivity, which are critical for their movement and genetic exchange.

Pesticide exposure represents another significant hazard, particularly the widespread use of neonicotinoids and other systemic agrochemicals. These substances have been shown to impair foraging behavior, navigation, learning, and reproduction in bees, even at sub-lethal doses. Chronic exposure leads to colony-level effects such as reduced brood development and increased mortality, contributing to colony collapse disorder in honey bees and population declines in wild species (Woodcock et al. 2016).

In addition to chemical threats, climate change is altering the spatial and temporal dynamics of plant-pollinator interactions. Shifts in temperature and precipitation patterns disrupt the synchronized timing or phenology between flowering plants and their pollinators, leading to mismatches that can reduce pollination success and plant reproductive output (Kudo and Ida 2013). These mismatches are especially problematic in ecosystems with short growing seasons or specialized plant-pollinator relationships.

Moreover, emerging diseases and invasive species exacerbate pollinator decline. Pathogens such as *Nosema* spp. and *Deformed Wing Virus*, often transmitted by the invasive *Varroa destructor* mite, have devastated managed honey bee colonies worldwide. These diseases can also spill over into wild bee populations, intensifying their decline (Goulson et al. 2015). Invasive species, including *Apis mellifera* in regions where it is not native, can outcompete indigenous pollinators for floral resources, alter community dynamics, and introduce novel pathogens.

Collectively, these threats are synergistic, meaning that their combined effects are often more detrimental than each factor alone, highlighting the urgent need for integrated pollinator conservation strategies.

4.7 Conservation Strategies

Safeguarding pollinators and their ecological services requires comprehensive, multidisciplinary conservation strategies that integrate ecological restoration, policy reforms, scientific research, and public engagement. Habitat restoration stands at the forefront of pollinator conservation.

By establishing and maintaining pollinator-friendly habitats such as wildflower meadows, hedgerows, and urban green spaces, ecologists aim to reintroduce native flowering plants, provide continuous floral resources, and ensure the availability of nesting sites for a wide range of pollinators. These restored landscapes not only support pollinator abundance and diversity but also enhance ecological connectivity across fragmented habitats (Dicks et al. 2016).

In tandem with ecological interventions, policy and regulation play a crucial role in mitigating human-induced pressures on pollinators. Many countries have initiated restrictions or bans on harmful pesticides like neonicotinoids, while promoting agroecological practices that reduce chemical dependency and encourage crop diversity. Policy frameworks such as the EU Pollinators Initiative underscore the importance of integrating biodiversity-friendly practices into mainstream agriculture (Vanbergen and the Insect Pollinators Initiative 2013). Moreover, financial incentives for farmers to maintain pollinator habitats such as flower-rich field margins and uncultivated buffer zones can align economic interests with ecological goals. Scientific research and long-term monitoring are indispensable for informing evidence-based conservation.

Monitoring programs help track population trends, identify emerging threats, and assess the effectiveness of conservation interventions. Studies focusing on pollinator behavior, disease dynamics, and plant-pollinator networks are crucial for designing adaptive management strategies that reflect ecological complexity (Potts et al. 2010).

Equally important is the role of public awareness and citizen participation. Educational campaigns, school-based programs, and citizen science initiatives have proven effective in fostering environmental stewardship. For example, programs like the Great Pollinator Project engage the public in pollinator surveys, creating valuable datasets while cultivating a culture of conservation. Grassroots efforts often lead to localized solutions, particularly in urban and agricultural landscapes, where changes in gardening, land use, and pesticide practices can significantly influence pollinator health. Ultimately, protecting pollinators demands collaborative action across disciplines and sectors, combining ecological knowledge with social engagement and policy innovation to build resilient ecosystems and secure global food production systems.

4.8 Discussion

Pollinators are indispensable to the stability and functioning of terrestrial ecosystems, playing a critical role in sustaining biodiversity, ensuring plant reproduction, and supporting global agricultural productivity. The intricate and co-evolved relationships between plants and their pollinators are vital for maintaining the ecological networks that underpin food webs and ecosystem services. With over 75% of global crop species dependent on animal-mediated pollination, the economic and nutritional importance of pollinators cannot be overstated. However, these essential organisms are increasingly threatened by habitat destruction, pesticide use, climate change, disease, and invasive species, which collectively compromise their populations and disrupt the delicate balance of plant-pollinator interactions. As the threats to pollinators intensify, their decline has far-reaching implications not only for ecosystems but also for food security and human livelihoods.

To safeguard pollinators and preserve their crucial ecological and economic services, urgent and integrated conservation strategies are required. These strategies should encompass habitat restoration, policy reforms, scientific research, and public engagement, ensuring that both natural and managed pollinator populations are supported and protected. Habitat restoration and the creation of pollinator-friendly landscapes are essential steps toward rebuilding connectivity and diversity in fragmented ecosystems.

Policy frameworks that regulate harmful pesticides and incentivize biodiversity-friendly agricultural practices will play a pivotal role in mitigating anthropogenic pressures. Furthermore, long-term scientific monitoring and research into pollinator health and behavior are necessary to adapt to emerging threats and evaluate conservation efforts. Public education and citizen science initiatives also have the potential to foster a greater understanding of pollinator ecology, thus encouraging collective action for their protection. Ultimately, the conservation of pollinators requires a collaborative, multi-disciplinary approach that integrates ecological knowledge, innovative policies, and active community involvement. By ensuring the health and abundance of pollinators, we can secure the future of global biodiversity, agricultural productivity, and food security, thus ensuring a sustainable and resilient world for generations to come.

References

Chittka L, Raine NE (2006) Recognition of flowers by pollinators. Curr Biol 16(17):R707–R710

Cronk QC, Ojeda I (2008) Bird-pollinated flowers in the tropics. Springer Science & Business Media

Dicks LV, Shen J, Jervis MA (2016) A global review of pollinator conservation. Conserv Biol 30(5):1026–1035

Eilers EJ, Kremen C, Greenleaf SS (2011) Contribution of pollinator-mediated crops to the global food supply. Nature 474(7350):87–90

Fenster CB, Armbruster WS, Wilson P, Dudash MR, Thomson JD (2004) Pollination syndromes and floral specialization. Annu Rev Ecol Evol Syst 35:375–403

Fleming TH, Geiselman CK, Kress WJ (2009) The evolution of bat pollination: a phylogenetic perspective. Ann Bot 104(4):1017–1033

Garibaldi LA, Steffan-Dewenter I, Kremen C, Morales JM, Bommarco R, Cunningham SA (2013) Wild pollinators enhance fruit set of crops regardless of honey bee abundance. Science 339(6127):1608–1611

Goulson D, Nicholls E, Botías C, Rotheray EL (2015) Bee declines driven by combined stress from parasites, pesticides, and lack of flowers. Science 347(6229):1255957

IPBES (2016) The assessment report on pollinators, pollination and food production. Intergovernmental Science-Policy Platform on Biodiversity and Ecosystem Services

Klein AM, Steffan-Dewenter I, Tscharntke T (2007a) Spatiotemporal dynamics of insect pollinator communities and pollination services in different landscapes. Oecologia 153(3):529–537

Klein AM, Vaissière BE, Cane JH, Steffan-Dewenter I, Tscharntke T (2007b) Importance of pollinators in changing landscapes for world crops. Proc R Soc B Biol Sci 274(1608):303–313

Kremen C, Williams NM, Aizen MA (2007) Pollination and other ecosystem services provided by insects. Insect Conserv Divers 1(4):125–131
Kudo G, Ida TY (2013) Effects of climate change on plant-pollinator interactions in Japan. Popul Ecol 55(4):441–451
Leonard AS, Masek P, Papaj DR (2011) The role of flower symmetry in shaping foraging behavior in bumblebees. Anim Behav 81(6):1203–1211
Memmott J, Craze PG, Waser NM (2004) Global warming and the disruption of plant-pollinator interactions. Ecol Lett 7(2):185–190
Ollerton J, Winfree R, Tarrant S (2011) How many flowering plants are pollinated by animals? Oikos 120(3):321–326
Potts SG, Imperatriz-Fonseca VL, Nguyen GH (2010) Global pollinator declines: the drivers and impacts. Science 331(6022):294–296
Raguso RA (2008) Wake up and smell the roses: the ecology and evolution of floral scent. Annu Rev Ecol Evol Syst 39:549–569
Thomson JD (2003) Pollen limitation of plant reproduction: ecological and evolutionary causes and consequences. In: Pollination ecology. Springer, Dordrecht, pp 60–82
van der Kooi CJ, Vallejo-Marín M, Leonhardt SD (2021) Mutualisms and (a)symmetry in plant-pollinator interactions. Curr Biol 31(2):R91–R99
van der Niet T, Johnson SD, Shuttleworth A (2011) Plant–pollinator coevolution and its implications for conservation in the Cape Floral Region of South Africa. Evol Ecol Res 13(8):765–786
Vanbergen AJ, the Insect Pollinators Initiative (2013) Threats to pollinators: a global review. Science 342(6167):1095–1098
Winfree R, Bartomeus I, Cariveau DP (2009) Native pollinators in anthropogenic habitats. Front Ecol Environ 7(4):1–8
Woodcock BA, McFrederick QS, Goulson D, Rotheray EL (2016) Pesticide contamination of the food web of the native bumblebee, *Bombus terrestris*, in a dynamic agricultural landscape. Sci Total Environ 578:1–8

Climate Change and Threat to Pollinators

5

Shyamasree Ghosh, Succandra Chowdhury,
and Md. Nurul Hasan

Abstract

Pollinators, including both managed species like *Apis mellifera* and wild pollinators such as bees, butterflies, birds, and bats, are vital for maintaining biodiversity, ecosystem resilience, and global food security. These organisms enable the reproduction of approximately 87% of flowering plants and are responsible for the pollination of 75% of major food crops. However, pollinator populations are facing unprecedented declines, with climate change identified as a significant and accelerating driver of this crisis. Rising temperatures, altered precipitation patterns, increased frequency of extreme weather events, and other climate-induced factors are disrupting the delicate relationships between pollinators and plants, leading to temporal and spatial mismatches, habitat loss, and reduced pollination efficiency. Additionally, climate change exacerbates the impacts of other anthropogenic stressors, including habitat fragmentation, pesticide exposure, and the spread of invasive species and pathogens. These disruptions have profound ecological and economic implications, as reduced pollination threatens agricultural productivity and food security, particularly in regions heavily reliant on pollinator-dependent crops. This chapter provides a comprehensive review of the multifaceted impacts of climate change on pollinator behavior, physiology, distribution, and interactions with plants. It also examines the cascading effects on plant-pollinator networks and outlines adaptive strategies for mitigating these

S. Ghosh (✉)
School of Biological Sciences, National Institute of Science Education and Research (NISER), Jatni, Odisha, India

S. Chowdhury
Post Graduate Department of Zoology, Bethune College, Kolkata, West Bengal, India

M. N. Hasan
Zoological Survey of India (Ministry of Environment, Forest and Climate Change, Government of India), Kolkata, West Bengal, India

S. Ghosh, D. Banerjee (eds.), *Pollinators: Life, Biology and Threats*,
https://doi.org/10.1007/978-981-95-0359-9_5

impacts, with a focus on conservation practices that enhance ecosystem resilience and ensure sustainable pollination services in a changing climate.

Keywords
Pollinators · Climate change · Biodiversity · Pollination networks · Ecosystem services · Food security

5.1 Introduction

Pollinators are indispensable components of terrestrial ecosystems, facilitating the reproduction of over 87% of flowering plant species globally and directly contributing to the productivity of approximately 75% of major food crops (Klein et al. 2007). These services are essential for the maintenance of biodiversity, ecosystem resilience, and agricultural output. Particularly, managed pollinators such as Apis mellifera (honey bees) and a diverse array of wild pollinators including bumblebees, solitary bees, butterflies, moths, birds, and bats play pivotal roles in the pollination of fruits, vegetables, nuts, and oilseeds (Garibaldi et al. 2013; Potts et al. 2016). The global economic value of insect pollination has been estimated at US$235–577 billion annually (IPBES 2016), emphasizing the critical interdependence between pollinator health and food security. However, mounting evidence indicates that pollinator populations are in sharp decline due to a combination of anthropogenic pressures, with climate change emerging as a predominant and accelerating threat (Goulson et al. 2015; Potts et al. 2010). Climate change is characterized by rising global temperatures, altered precipitation regimes, more frequent and intense extreme weather events, and elevated atmospheric CO_2 levels, each of which can disrupt ecological relationships, particularly those between pollinators and the plants they service (Scaven and Rafferty 2013; IPCC 2021). These disruptions manifest through multiple pathways, including phenological mismatches between pollinators and flowering plants, shifts in geographic ranges, changes in foraging behavior, reductions in floral abundance and nutritional quality, and increased vulnerability to pests, diseases, and invasive species (Memmott et al. 2007; Kerr et al. 2015; Gérard et al. 2020). Phenological shifts, where the timing of biological events is altered, are particularly critical in pollination systems. For instance, warming temperatures may cause earlier flowering in plants, while pollinators such as bees may not advance their emergence at the same rate, leading to temporal mismatches that reduce pollination success (Forrest 2015; Kudo and Ida 2013). Geographic range shifts, where pollinators move to higher altitudes or latitudes in response to thermal stress, may isolate them from co-evolved plant species and disrupt mutualistic interactions (Rafferty et al. 2015; Bartomeus et al. 2011). Additionally, increased climatic variability may reduce pollinator activity during critical foraging periods and increase mortality during overwintering phases (Radchuk et al. 2013). Climate change also interacts synergistically with other anthropogenic stressors, such as habitat loss, agricultural intensification, pesticide exposure, the spread of genetically modified organisms, and pathogen spillover from managed to wild populations (Goulson et al. 2015). For example, sublethal pesticide exposure has been shown to impair thermoregulation, navigation, and immune function in bees, making them less resilient to climate-induced stress (Pisa et al. 2015). Meanwhile, habitat fragmentation and monoculture farming reduce floral diversity and nesting resources, compounding the challenges pollinators face in a rapidly changing environment (Kennedy et al. 2013; Roulston and Goodell 2011). Pollinator decline under climate stress not only threatens wild biodiversity but also undermines agricultural stability and human nutrition. Crops that depend on animal pollination tend to be high-value and nutrient-rich, including fruits, vegetables, and nuts, which provide essential micronutrients like vitamins A and C, calcium, and folate (Eilers et al. 2011). Consequently, a reduction in pollinator abundance and efficiency can lead to decreased yields and lower quality of these foods, disproportionately affecting smallholder farmers and populations in regions where food security is already precarious (Smith et al. 2015; Gallai et al. 2009). Given the interconnectedness of climate dynamics and pollinator health, it is crucial to develop a comprehensive understanding of how climate change alters pollinator behavior, physiology, community composition, and plant-pollinator networks. Such understanding is essential for guiding policy, conservation planning, and the design of climate-resilient agricultural systems. This chapter will systematically examine the multifactorial threats posed by climate change to pollinators, the ecological and economic consequences of their decline, and the adaptive strategies necessary to ensure the sustainability of pollination services in the Anthropocene.

5.2 Climate Change and Impact on Pollinators in the Global Scenario

Pollinators are critical providers of ecosystem services that sustain both agricultural productivity and biodiversity. They facilitate the reproduction of countless wild and cultivated plant species, thereby playing an integral role in ecosystem functioning, food security, and rural livelihoods (IPBES 2016; Potts et al. 2016). However, mounting evidence points to a concerning global decline in pollinator diversity and abundance, largely driven by a complex interplay of anthropogenic stressors. Among these, climate change stands out as a particularly pervasive and multifactorial threat, compounding the effects of habitat fragmentation, pesticide exposure, invasive species, and disease dynamics (Goulson et al. 2015). Rising global temperatures, shifting precipitation patterns, and increased frequency of extreme climatic events are disrupting the geographic ranges, phenologies, and population dynamics of numerous pollinator taxa. Species adapted to specific microclimates, such as those inhabiting montane, arctic, or Mediterranean environments, are especially vulnerable, as their narrow ecological niches leave little room for adaptive migration (Rafferty and Ives 2011; Inouye 2020). In alpine ecosystems, for instance, warming temperatures have led to the upward migration of both plants and pollinators, intensifying competition and displacing endemic species unable to relocate or adapt in time (Hoiss et al. 2012; Classen et al. 2015). Pollinator range shifts often lead to mismatches in plant-pollinator interactions, particularly when the phenological responses of plants and their pollinators to climate cues differ in magnitude or direction. Such mismatches can reduce reproductive success in plants and limit foraging resources for pollinators, potentially leading to cascading ecological consequences (Memmott et al. 2007; Kudo and Ida 2013). Additionally, as species shift their ranges in response to climate stress, novel interspecific interactions may arise, including the arrival of invasive species or new competitors that further destabilize native pollinator communities (Brown et al. 2016; Carrasco et al. 2021). Empirical and modeling studies have reinforced the global scope of these impacts. For example, a comprehensive modeling analysis using species distribution models (SDMs) on pollinators across Brazil, a biodiversity hotspot, projected that climate change could significantly alter the distributions of key pollinator species, thereby affecting crop yield, total agricultural production value, and regional food security (Giannini et al. 2017). These disruptions are not isolated to the tropics; similar predictions have been made for temperate regions, where bumblebee populations are contracting southward in Europe and North America due to thermal limitations and habitat saturation at higher latitudes (Kerr et al. 2015). Moreover, pollinator decline under climate stress is often exacerbated by land-use changes such as deforestation, agricultural intensification, and urban expansion, which reduce floral diversity and nesting habitats (Kennedy et al. 2013; Roulston and Goodell 2011). This habitat degradation further limits the ability of pollinators to adapt to shifting climate envelopes. In many cases, climate change also modifies the epidemiology of pollinator pathogens, expanding the range and virulence of viruses, fungi, and parasites that afflict both wild and managed pollinators (Cavigli et al. 2016; Manley et al. 2015). Given the global extent and multifaceted nature of these climate-driven threats, there is an urgent need to incorporate pollinator conservation into broader climate adaptation and agricultural planning frameworks. Strategies must emphasize landscape connectivity, floral resource diversity, and the resilience of ecological networks to ensure the continued delivery of pollination services under future climate scenarios.

5.3 Species Phenology and Climate Change

The timing of biological events, or phenology, is highly sensitive to climate change, and its effects are particularly evident in plant-pollinator interactions. These interactions are integral to ecosystem functioning, yet climate-induced phenological shifts can disrupt the timing of these interactions, especially when species respond to climate cues at different rates. Such mismatches can destabilize ecosystems and negatively impact the survival of both plants and their pollinators. Research has shown that, in response to warming temperatures, plants tend to advance their flowering times, whereas insect pollinators may not adjust at the same pace. A notable study conducted in Germany, which tracked phenological changes since the 1980s, revealed that plant species exhibited more pronounced shifts in their blooming times than many insect pollinators. This misalignment between plant and pollinator activity increases the risk of mismatched interactions, leading to reduced pollination success and potentially lower reproductive outcomes for plants (Freimuth 2022). Such desynchronization between plant and pollinator activity can have long-term consequences, including declines in pollination success and the availability of resources for pollinators. These disruptions could lead to a reduction in both plant and pollinator populations, threatening biodiversity and undermining ecosystem resilience. Over time, this could result in cascading effects, influencing food webs and the overall health of ecosystems.

5.4 Impact on Plant-Pollinator Interaction

The mutualistic relationship between flowering plants and their pollinators forms the cornerstone of terrestrial biodiversity and food security. This dynamic interaction is highly sensitive to environmental fluctuations, and climate change is now widely recognized as a key driver of disruption in these relationships. By influencing variables such as temperature, precipitation, atmospheric CO_2 concentration, and the frequency of extreme weather events, climate change alters critical traits in both plants and pollinators, including phenology, physiology, spatial distribution, and inter-species dependencies (Kearns et al. 1998; Potts et al. 2010; Rafferty and Ives 2011). One of the most profound impacts of climate change on plant-pollinator systems is the disruption of temporal synchrony. Plants and pollinators are finely attuned to seasonal cues flowering must coincide with the active foraging period of pollinators for optimal reproductive success. However, rising temperatures have been shown to advance plant flowering times more rapidly than the emergence times of insect pollinators, resulting in phenological mismatches (Forrest 2015; Fitter and Fitter 2002; Kudo and Ida 2013). These mismatches reduce pollination efficiency, especially for specialized plant species that depend on a narrow range of pollinators (Memmott et al. 2007; Hegland et al. 2009). Beyond timing, plant physiological traits are also being altered. Warmer temperatures and altered rainfall patterns influence plant metabolism, including the production of nectar and pollen, which are vital resources for pollinators. For instance, elevated temperatures have been observed to reduce nectar volume and change its sugar composition, making it less attractive or nutritionally insufficient for pollinators (Hoover et al. 2012; Takkis et al. 2015). A study by Giejsztowt et al. (2020) demonstrated that warming-induced reductions in nectar quality and quantity not only decrease pollinator visitation rates but also influence the composition of the pollinator community, favoring generalists over specialists. Furthermore, interactions with co-occurring plant species, including invasives, are shifting under climate change. Altered flowering phenology among native and invasive species can modify floral competition, leading to a decrease in pollinator visits to native plants (Gibson et al. 2012). This competitive imbalance, coupled with increased climatic stress on native flora, can diminish reproductive success and threaten biodiversity. Systemic stress extends to ecosystem-wide interactions, as climate change also alters the dynamics between plants, pollinators, pathogens, and herbivores. Borghi et al. (2019) emphasized that global warming influences the biosynthesis of floral volatiles and defense metabolites in plants, which are essential for attracting pollinators and resisting pests. These changes may compromise floral signaling, further disrupting pollination services. Additionally, pathogens and parasites that affect both plants and pollinators, such as fungal infections in flowers or *Nosema* in bees, are spreading due to climate variability, compounding the threat (Le Conte and Navajas 2008; Goulson et al. 2015). The cumulative impact of these shifts extends beyond ecological consequences to directly threaten agricultural productivity and food security. Pollinator-dependent crops such as apples, almonds, berries, and coffee are particularly vulnerable to reduced pollination services under climate stress (Klein et al. 2007; Ricketts et al. 2008). The loss of pollination can result in lower yields, reduced crop quality, and economic losses, especially in regions highly reliant on pollinator services. Ultimately, these disruptions reflect a broader erosion of ecosystem resilience. The fragmentation and collapse of plant-pollinator networks weaken the capacity of ecosystems to adapt to changing conditions, increasing their vulnerability to additional stresses like habitat loss, pesticide exposure, and invasive species (Tylianakis et al. 2008; González-Varo et al. 2013). This underscores the urgent need for integrated conservation strategies that account for the multifaceted challenges posed by climate change to both plants and their pollinators.

5.5 Impact on Bumblebees

Bumblebees (*Bombus* spp.) are essential pollinators found across diverse ecosystems, ranging from arctic tundra to temperate and tropical environments. Their robust morphology and buzz pollination ability make them invaluable for the pollination of both wild flora and agricultural crops such as tomatoes, blueberries, and clover (Goulson 2010; Rader et al. 2016). However, as cold-adapted and eusocial insects, bumblebees are particularly sensitive to environmental disturbances, especially those induced by climate change. Rising temperatures have been directly linked to significant declines in bumblebee abundance and range. A continent-wide study in North America revealed that bumblebee presence is more closely associated with temperature anomalies than with floral resource availability or precipitation variability (Jackson et al. 2022). These findings indicate that climate-driven thermal stress is a primary factor influencing bumblebee occupancy. Although some species exhibit physiological adaptations, such as modifications in metabolic traits or seasonal activity patterns, these responses vary widely among taxa. For instance, in Fiji, the endemic bee *Homalictus fijiensis* was found to have elevated metabolic rates but lower thermal tolerance compared to invasive species like *Braunsapis puangensis* and *Ceratina dentipes*, which are more heat-resilient and ecologically competitive under warming scenarios (da Silva et al. 2021). Compounding these physiological vulnerabilities are other climate-exacerbated threats, including habitat fragmentation, pesticide exposure, invasive species, and the proliferation of parasites and pathogens. The interaction between rising temperatures and disease dynamics—such as increased susceptibility to *Nosema* or *Crithidia* infections—has further destabilized bumblebee populations (Cameron and Sadd 2020; Maebe et al. 2021). While some species display plastic responses to environmental shifts, the overall trend remains one of reduced biodiversity and pollination service loss.

5.6 Climate Change as a Stress Multiplier

Beyond direct temperature and precipitation changes, climate change acts as a multifaceted ecological stressor by amplifying biological threats and altering ecological processes. It facilitates the spread of pests, enhances competition from invasive species, degrades soil microbial communities, and influences plant physiology in ways that affect both pollinators and plants (Shahzad et al. 2021). The redistribution of species under shifting climate regimes has further disrupted ecological synchrony. As temperature bands move poleward or upward in elevation, organisms adjust their ranges accordingly, often resulting in spatial mismatches between interdependent species. A compelling case is the mutualism between the migratory bat *Leptonycteris nivalis* and *Agave* plants. Climate-induced shifts in suitable habitat for agaves have restricted nectar resources for these bats, threatening their migratory patterns and reproductive success (Gómez-Ruiz and Lacher 2019). Additionally, climate stress alters the production of plant volatile organic compounds (VOCs), particularly terpenoids, which are critical for pollinator attraction and plant defense. These secondary metabolites are influenced by environmental stressors and play key roles in plant resilience and the mediation of plant-pollinator communication (Abbas et al. 2022). Reductions or compositional changes in VOCs can undermine pollination success and further destabilize ecosystem functioning. Agricultural systems are also vulnerable. Crops that rely heavily on pollinators, such as coffee, are predicted to experience significant yield declines due to shrinking climate-suitable zones and declining pollinator diversity. Adaptive management practices, including habitat diversification, shade-grown systems, and conservation of wild pollinators, are critical to mitigate these impacts (Imbach et al. 2017).

5.7 Impact on Species Distribution and Pollination Networks

Climate change is reshaping species distributions globally, with significant repercussions for ecological interactions, particularly pollination networks. As species shift their ranges in response to climate cues, pre-existing mutualisms may be severed, and new, untested relationships may form. These spatial rearrangements can destabilize plant-pollinator dynamics, reduce pollination efficiency, and threaten reproductive success in both wild and cultivated species. The disruption of the *Leptonycteris nivalis-Agave* mutualism offers a prominent example of how geographic shifts can undermine ecological dependencies (Gómez-Ruiz and Lacher 2019). Likewise, insect pollinators are affected through changes in voltinism (the number of generations per year), emergence timing, and overall abundance. These life-history shifts, when decoupled from host plant phenology, can result in either missed interactions or heightened pest pressure (Eigenbrode et al. 2022). Temperate regions, especially in the Northern Hemisphere, have reported sharp declines in pollinator diversity due to rising temperatures and habitat degradation (Vasiliev and Greenwood 2021). The weakening of plant-pollinator interactions in these systems has been associated with lower rates of successful seed and fruit production (Byers and Chang 2017). Another underappreciated mechanism by which climate affects pollination is through its influence on plant secondary metabolism. As mentioned, VOCs not only serve as attractants for pollinators but also mediate interspecies competition and defense. Changes in VOC profiles due to warming or drought stress can reduce the effectiveness of pollinator attraction and interfere with plant reproductive success (Abbas et al. 2022). Overall, the redistribution of species, combined with physiological, behavioral, and metabolic responses to climate stress, is leading to increasingly fragile pollination systems. These disruptions highlight the urgent need for integrative conservation strategies that include assisted migration, landscape connectivity, and ecosystem-based adaptation approaches.

5.8 Conclusion

Climate change poses a profound and escalating threat to global pollinator communities, the mutualistic networks they support, and the vital ecosystem services they provide. Through mechanisms such as altered phenology, geographic range shifts, disrupted plant-pollinator synchrony, and increased exposure to pests, pathogens, and pollutants, climate change acts not as a singular stressor but as a complex and compounding ecological force. Pollinators, both wild and managed, are already showing signs of decline across diverse biogeographic regions from tropical ecosystems to temperate and alpine zones, with cascading consequences for biodiversity, food security, and human livelihoods. The erosion of plant pollinator interactions, especially in climate-sensitive systems, threatens not only the reproductive success of numerous flowering plants but also the yield, nutritional quality, and stability of pollinator-dependent crops. Moreover, the degradation of floral resources, shifts in volatile organic compound (VOC) profiles, and climate-mediated breakdown of ecological synchrony further accentuate the fragility of these interactions in an increasingly unpredictable world. Species-specific vulnerabilities, as seen in bumblebees and specialized pollinators like the migratory bat *Leptonycteris nivalis*, underscore the urgent need for targeted conservation interventions. Empirical studies and predictive models alike warn that without immediate action, many pollinator species may face range contractions, local extinctions, or loss of ecological function. In response, it is imperative to integrate pollinator conservation into global climate adaptation strategies, agricultural planning, and biodiversity policies. Approaches such as enhancing landscape connectivity, protecting and restoring floral and nesting habitats, promoting agroecological practices, and supporting climate-resilient pollinator species must become central components of environmental governance. Furthermore, transdisciplinary research and policy coordination at local, national, and international levels are essential to develop adaptive, evidence-based solutions that safeguard pollination services under future climate scenarios. In the Anthropocene, where human-driven climate change is reshaping life on Earth, protecting pollinators is not only a matter of ecological stewardship but also of ensuring long-term food sovereignty, ecosystem resilience, and planetary health.

References

Abbas F, O'Neill Rothenberg D, Zhou Y, Ke Y, Wang HC (2022) Volatile organic compounds as mediators of plant communication and adaptation to climate change. Physiol Plant 174(6):e13840

Bartomeus I et al (2011) Geographic range shifts may isolate pollinators from co-evolved plant species. Ecol Lett 14(3):219–227

Borghi M et al (2019) Global warming influences the biosynthesis of floral volatiles and defense metabolites in plants. New Phytol 223(2):681–691

Brown MJF et al (2016) Novel interspecific interactions arising from climate change are destabilizing native pollinator communities. Glob Chang Biol 22(3):1208–1218

Byers D, Chang C (2017) Climate change and pollinator decline: fragmentation of plant-pollinator networks. Ecol Appl 27(1):202–212

Cameron SA, Sadd BM (2020) Increased susceptibility to Nosema and Crithidia infections in bumblebees under climate stress. Insect Sci 27(6):1134–1145

Carrasco D et al (2021) Climate-induced range shifts and the introduction of new species alter pollinator communities. Ecol Lett 24(6):1182–1193

Cavigli D et al (2016) Climate change modifies the epidemiology of pollinator pathogens. Proc R Soc B Biol Sci 283(1836):20160534

Classen AT et al (2015) Warming-induced shifts in species distributions affect ecosystem functioning. Nature 521(7550):139–142

Da Silva MA et al (2021) Thermal tolerance of endemic versus invasive species: comparative study in Fiji. Ecol Entomol 46(2):286–297

Eigenbrode SD et al (2022) Changes in voltinism and abundance in pollinators under climate stress. Ecology 103(4):e3489

Eilers EJ et al (2011) Crop pollination services provided by native bees in the United States. Proc Natl Acad Sci 108(1):66–71

Fitter AH, Fitter RSR (2002) Rapid changes in flowering time and phenology in response to climate change. Science 296(5573):1689–1691

Forrest JRK (2015) Phenological mismatches between pollinators and flowering plants. Curr Opin Insect Sci 11:60–64

Freimuth J (2022) Phenological shifts in plant-pollinator interactions in response to climate change. J Ecol Res 35(2):121–134

Gallai N et al (2009) Economic valuation of the vulnerability of world agriculture confronted with pollinator decline. Ecol Econ 68(3):810–821

Garibaldi LA et al (2013) Wild pollinators enhance fruit set of crops irrespective of honey bee abundance. Science 339(6127):1608–1611

Gérard M et al (2020) Increased vulnerability of plant-pollinator networks to climate-induced disruptions. Glob Chang Biol 26(6):3010–3023

Giannini TC et al (2017) Impact of climate change on pollinator distributions and crop yields in Brazil. Ecol Lett 20(6):755–764

Gibson MR, Richardson DM, Pauw A (2012) Can floral traits predict an invasive plant's impact on native plant–pollinator communities? J Ecol 100(5):1216–1223. https://doi.org/10.1111/j.1365-2745.2012.02004.x

Giejsztowt D et al (2020) Warming-induced reductions in nectar quality and quantity decrease pollinator visitation rates. Oecologia 192(2):455–465

Gómez-Ruiz EP, Lacher TE Jr (2019) Climate change, range shifts, and the disruption of a pollinator–plant complex. Sci Rep 9:14048. https://doi.org/10.1038/s41598-019-50059-6

González-Varo JP et al (2013) Pollinator disruption under climate change and land-use change. J Ecol 101(3):460–471

Goulson D (2010) Bumblebees: their behaviour and ecology. Oxford University Press, Oxford

Goulson D et al (2015) Declining pollinator populations and their consequences for global food security. Science 347(6229):1255957

Hegland SJ, Nielsen A, Lázaro A, Bjerknes A-L, Totland Ø (2009) How does climate warming affect plant–pollinator interactions? Ecol Lett 12(2):184–195. https://doi.org/10.1111/j.1461-0248.2008.01269.x

Hoiss B et al (2012) The impact of climate change on the distribution of plants and pollinators in temperate and alpine ecosystems. Ecol Lett 15(7):755–764

Hoover SER et al (2012) How warming affects the nectar composition of plants and its effect on pollinators. Funct Ecol 26(1):3–13

Imbach P et al (2017) Adaptive management practices for climate change in coffee agriculture. Agric Ecosyst Environ 248:215–222

Inouye DW (2020) The effects of climate change on plant-pollinator relationships in alpine ecosystems. Glob Ecol Biogeogr 29(3):514–525

IPBES (2016) The assessment report on pollinators, pollination and food production. Intergovernmental Science-Policy Platform on Biodiversity and Ecosystem Services

IPCC (2021) Climate change 2021: the physical science basis. Intergovernmental Panel on Climate Change

Jackson ML et al (2022) Thermal stress as a primary driver of bumblebee distribution and abundance. Ecol Appl 32(3):e02467

Kearns CA et al (1998) The role of pollinators in the reproductive success of flowering plants. Ecol Monogr 68(4):213–233

Kennedy CM et al (2013) The impact of agricultural intensification on the diversity of flower-visiting insects in North America. Proc Natl Acad Sci 110(3):1–6

Kerr JT et al (2015) Bumblebee populations in the northern hemisphere have been severely impacted by climate change. Nat Commun 6:10760

Klein A-M et al (2007) Pollination of crops and wild plants by insects. Science 313(5791):351–354

Kudo G, Ida TY (2013) Effects of climate change on the phenology of plants and pollinators in alpine ecosystems. Ecology 94(1):56–68

Le Conte Y, Navajas M (2008) Climate change: Impact on honey bee populations and diseases. Rev Sci Tech - Off int épizoot 27(2):485–497

Maebe K et al (2021) Temperature changes and the increase in disease susceptibility in bumblebees. J Invertebr Pathol 174:107442

Manley R et al (2015) The effects of climate-induced range expansion of pathogens on wild pollinators. Ecol Appl 25(2):536–546

Memmott J et al (2007) Global warming and the disruption of plant-pollinator networks. Science 316(5832):1767–1770

Pisa LW et al (2015) The impacts of sublethal pesticide exposure on pollinators: effects on thermoregulation, navigation, and immune function in bees. Environ Toxicol Chem 34(4):950–957

Potts SG et al (2010) Global pollinator declines: trends, impacts, and drivers. Trends Ecol Evol 25(6):345–353

Potts SG et al (2016) The role of pollinators in food security and ecosystem services. Food Secur 8(3):529–542

Radchuk V et al (2013) Impacts of climatic extremes on pollinators. Glob Chang Biol 19(8):2423–2432

Rader R et al (2016) Bumblebees as effective pollinators of agricultural crops. J Econ Entomol 109(4):1483–1493

Rafferty NE, Ives AR (2011) Effects of experimental shifts in flowering phenology on plant–pollinator interactions. Ecol Lett 14(1):69–74. https://doi.org/10.1111/j.1461-0248.2010.01557.x

Rafferty NE et al (2015) Shifting species distributions and climate change: a case study on pollinators. Glob Chang Biol 21(5):1510–1519

Ricketts TH, Regetz J, Steffan-Dewenter I, Cunningham SA, Kremen C, Bogdanski A, … Viana BF (2008) Landscape effects on crop pollination services: Are there general patterns? Ecol Lett 11(5):499–515. https://doi.org/10.1111/j.1461-0248.2008.01157.x

Roulston TH, Goodell K (2011) The effects of habitat fragmentation on the diversity of pollinators. Environ Entomol 40(5):1041–1049

Scaven VL, Rafferty NE (2013) Physiological effects of climate warming on flowering plants and insect pollinators and potential consequences for their interactions. Curr Zool 59(3):418-426

Shahzad A, Ullah S, Dar AA, Sardar MF, Mehmood T, Tufail MA, Shakoor A, Haris M (2021) Nexus on climate change: Agriculture and possible solution to cope future climate change stresses. Environ Sci Pollut Res Int 28(12):14211–14232. https://doi.org/10.1007/s11356-021-12649-8

Smith MD, Johnson AL, Thompson LJ (2015) Pollinator diversity and crop yield in agricultural systems. Ecol Lett 18(6):582–589. https://doi.org/10.1111/ele.12435

Takkis K, Tscheulin T, Tsalkatis P, Petanidou T (2015) Climate change reduces nectar secretion in two common Mediterranean plants. AoB Plants 7:plv111. https://doi.org/10.1093/aobpla/plv111

Tylianakis JM, Didham RK, Bascompte J, Wardle DA (2008) Global change and species interactions in terrestrial ecosystems. Ecol Lett 11(12):1351–1363. https://doi.org/10.1111/j.1461-0248.2008.01250.x

Vasiliev D, Greenwood S (2021) The role of climate change in pollinator decline across the Northern Hemisphere is underestimated. Sci Total Environ 775:145788. https://doi.org/10.1016/j.scitotenv.2021.145788

Insect Immunity: A Focus on Honey Bees

6

Shyamasree Ghosh and Md. Nurul Hasan

Abstract

The immune system in insects represents a highly evolved and dynamic suite of defenses that are pivotal for their survival in the face of diverse pathogens, including bacteria, viruses, fungi, and parasites. Insects rely solely on innate immunity, lacking the adaptive immune responses found in vertebrates. Within this domain, the honey bee (*Apis mellifera*) emerges not only as an ecologically and economically significant pollinator but also as a critical model for studying insect immunity due to its complex social structure and unique exposure to environmental and pathogenic stressors. This chapter provides an in-depth examination of the molecular and cellular components of insect immunity, focusing particularly on antiviral defenses in honey bees, the roles of immune signaling pathways such as Toll, Imd, Jak/STAT, and RNA interference (RNAi), and how environmental stressors like pesticides and malnutrition compromise immune function. Comparative studies with model organisms such as *Drosophila melanogaster* and *Anopheles gambiae* provide further insights into the evolutionary dynamics of immune specialization. Moreover, the COVID-19 pandemic brought renewed attention to viral ecology, indirectly affecting bee research and health through altered environmental conditions and research priorities. The chapter concludes by emphasizing the need for integrative approaches that combine molecular immunology, ecological monitoring, and policy reforms to safeguard pollinators in an era of escalating environmental challenges.

S. Ghosh (✉)
School of Biological Sciences, National Institute of Science Education and Research (NISER), Jatni, Odisha, India

M. N. Hasan
Zoological Survey of India (Ministry of Environment, Forest and Climate Change, Government of India), Kolkata, West Bengal, India

S. Ghosh, D. Banerjee (eds.), *Pollinators: Life, Biology and Threats*,
https://doi.org/10.1007/978-981-95-0359-9_6

Keywords

Insect immunity · Honey bees · Pollination · Antiviral defense · Immune suppression

6.1 Introduction

Insects have persisted as the most diverse and abundant group of animals on Earth, in part due to their effective innate immune systems. Insect immunity, unlike that of vertebrates, does not possess antigen-specific memory or antibody production. Instead, it relies on a two-tiered innate immune architecture composed of cellular and humoral responses that are rapidly activated upon detection of pathogens (Lemaitre and Hoffmann 2007). Honey bees (*Apis mellifera*) particularly provide a unique lens through which to study immunity, as their eusocial behavior contributes an additional layer of communal or "social immunity," where individual actions collectively reduce disease prevalence at the colony level (Cremer et al. 2007; Evans and Spivak 2010). These behaviors include allogrooming, removal of diseased brood (hygienic behavior), propolis deposition, and altered foraging behaviors to reduce pathogen exposure. Despite their sophisticated immune systems and communal defenses, honey bees are increasingly threatened by multifactorial stressors such as viral epidemics, pesticide exposure, monocultural diets, and habitat fragmentation. Pathogens like *Nosema ceranae*, *Varroa destructor*-vectored viruses, and environmental toxins interact synergistically, often overwhelming the bees' immune defenses and leading to phenomena such as colony collapse disorder (CCD) (Goulson et al. 2015; Genersch 2010). This has far-reaching consequences not only for ecosystem health, given the bees' role in pollination, but also for agricultural productivity and food security. It is thus imperative to deepen our understanding of the intricate interplay between honey bee immunity and environmental pressures to formulate informed conservation strategies.

6.2 Mechanisms of Insect Immunity

The insect immune system comprises three major defense components: physical barriers, cellular immunity, and humoral responses. The first line of defense includes the exoskeleton (cuticle) and the peritrophic membrane in the midgut, which mechanically obstructs pathogen entry (Vallet-Gely et al. 2008). Once these barriers are breached, cellular immunity takes over, mediated primarily by hemocytes. These cells perform phagocytosis of microbial invaders, encapsulation of larger parasites, and nodulation—a collective hemocyte response to foreign bodies (Lavine and Strand 2022). Humoral responses involve the secretion of antimicrobial peptides (AMPs), prophenoloxidase activation for melanization, and production of reactive oxygen species. AMPs such as defensins, abaecin, and apidaecin are synthesized upon immune challenge and act directly on microbial membranes (Casteels-Josson et al. 1994). The phenoloxidase cascade plays a crucial role in wound healing and encapsulation, contributing to pathogen sequestration and melanotic encapsulation. Several highly conserved immune signaling pathways regulate these responses. The Toll pathway, activated primarily by Gram-positive bacteria and fungi, leads to the transcription of AMPs through nuclear factor kappa-light-chain-enhancer of activated B cells (NF-κB) homologs such as Dorsal and Dif (Lemaitre et al. 1996). The Imd pathway responds to Gram-negative bacteria via peptidoglycan recognition proteins (PGRPs) and culminates in AMP production through the transcription factor Relish (Myllymäki et al. 2014). The Jak/STAT pathway, first discovered in *Drosophila*, is involved in antiviral and antiparasitic responses and regulates hematopoiesis and tissue regeneration (Agaisse and Perrimon 2004). The RNA interference (RNAi) pathway, especially critical in honey bees, provides sequence-specific silencing of viral genomes, making it the principal mechanism against RNA viruses (Brutscher et al. 2015). These pathways operate both independently and synergistically, ensuring a multifaceted and energy-efficient response to diverse microbial threats.

6.3 Antiviral Mechanisms in Honey Bees

Honey bees are host to a broad array of viruses, many of which are exacerbated by the presence of the *Varroa destructor* mite. These include Deformed Wing Virus (DWV), Israeli Acute Paralysis Virus (IAPV), Kashmir Bee Virus (KBV), and Black Queen Cell Virus (BQCV), among others. DWV has particularly emerged as a central agent of colony declines, facilitated by *Varroa*'s ability to bypass physical barriers and directly inject viral particles into bee hemolymph (Rosenkranz et al. 2010). The primary line of antiviral defense in honey bees is the RNAi pathway. Viral double-stranded RNA is recognized by Dicer-like enzymes, cleaved into small interfering RNAs (siRNAs), and incorporated into the RNA-induced silencing complex (RISC), which then targets complementary viral RNA for degradation (Maori et al. 2009; Brutscher et al. 2017). Experimental inoculations with synthetic dsRNA have demonstrated significant reductions in viral titers, underlining the pathway's therapeutic potential (Hunter et al. 2010). Besides RNAi, evidence suggests the involvement of the Toll and Jak/STAT pathways in antiviral immunity. Transcriptomic studies reveal upregulation of AMPs and stress-response genes following viral challenge (Evans et al. 2006). However, unlike *Drosophila*, the honey bee genome encodes fewer immune genes, potentially due to social buffering mechanisms (Evans et al. 2006). This streamlined immune repertoire may render bees more susceptible to highly virulent pathogens under stress, indicating the need for environmental conditions that support both immune competence and social cohesion.

6.4 Environmental Stressors and Immune Function

The immune health of honey bees is highly susceptible to environmental variables, many of which act as immunosuppressants. Pesticides, especially neonicotinoids such as imidacloprid and clothianidin, are known to impair the expression of immune genes, including defensin and abaecin, weaken hemocyte function, and increase susceptibility to viral infections (Di Prisco et al. 2013; Brandt et al. 2017). Sublethal exposure compromises the Toll and Imd pathways, allowing viral proliferation and reduced longevity. Nutritional stress arising from monocultures, habitat loss, or seasonal scarcity leads to reduced levels of essential amino acids, lipids, and micronutrients that are crucial for AMP synthesis and detoxification enzymes (Alaux et al. 2010). Protein-poor diets lower hemocyte counts and downregulate immune gene expression. Additionally, pollen diversity directly correlates with immune resilience, reinforcing the need for floral heterogeneity in bee landscapes. Climate change imposes another major burden. Elevated temperatures can dysregulate pathogen replication rates and interfere with developmental thresholds in bees, while extreme weather events limit foraging opportunities and impair nutritional intake (Le Conte and Navajas 2008). The combined impact of these stressors often results in synergistic effects that overwhelm even the best-adapted colonies, amplifying mortality and collapse rates across regions.

6.5 Comparative Immunity—Insights from Model Organisms

Comparative studies involving *Drosophila melanogaster*, *Anopheles gambiae*, and other model insects have illuminated both conserved and divergent aspects of insect immunity. *Drosophila*, for instance, possesses an expansive catalog of AMPs and immune regulators that have served as templates for understanding Toll and Imd signaling (Lemaitre and Hoffmann 2007). *Anopheles gambiae* adds further depth through its TEP-mediated immune response and complement-like system used in *Plasmodium* resistance (Blandin et al. 2004). Honey bees, in contrast, exhibit a reduced number of immune genes, possibly reflecting a reliance on behavioral defenses and social mechanisms for disease mitigation (Evans et al. 2006). For example, hygienic behavior where bees detect and remove infected brood serves as a colony-level immune strategy that limits pathogen load without invoking costly physiological responses (Spivak and Gilliam 1998). Such comparisons underscore the unique evolutionary paths taken by solitary versus eusocial insects and reveal how collective behavior can supplement or even replace certain immune functions at the individual level.

6.6 The Impact of COVID-19 on Bee Research and Health

The COVID-19 pandemic, while primarily a human health crisis, had significant ripple effects on pollinator research, conservation, and health. Global lockdowns disrupted regular fieldwork and monitoring activities, leading to gaps in long-term data collection (Dicks et al. 2021). Conversely, the pandemic's focus on viral transmission heightened interest in bee virology and immunity, fostering new funding streams and cross-disciplinary research into viral ecology and immunity (Rzymski et al. 2021). Reduced industrial activity during lockdowns temporarily improved air quality and lessened agrochemical application in certain areas, which may have incidentally benefited pollinators. However, supply chain disruptions impacted beekeeping practices, including delayed queen shipments and reduced access to mite treatments. These mixed outcomes highlight the need for resilient, decentralized research and conservation infrastructures that can adapt to global emergencies while maintaining pollinator health initiatives.

6.7 Hygienic Behavior as a Key Immunological Strategy

An important immunological behavior in honey bees is hygienic behavior, which serves as a critical form of social immunity within the colony. Hygienic behavior refers to the ability of worker bees to detect and remove infected larvae and pupae, which significantly reduces the spread of pathogens like *Nosema ceranae* and certain viruses (Spivak and Gilliam 1998). This behavior relies on the workers' ability to discern diseased individuals by detecting abnormal smells or physical cues, often linked to the presence of pathogens. Worker bees use olfactory cues to recognize diseased brood, which can be removed before the infection spreads to other members of the colony (Spivak et al. 2003). It has been shown that colonies with high levels of hygienic behavior have a lower prevalence of diseases such as foulbrood and *Nosema*, demonstrating that social immunity can greatly complement individual immunity. This collective immune behavior is particularly crucial in the context of honey bee viruses, which can spread quickly through large colonies. Studies have found that colonies exhibiting better hygienic behavior can maintain higher colony survival rates even in the face of persistent viral infections. This ability of honey bee colonies to prevent pathogen transmission through coordinated behaviors highlights the synergistic relationship between individual immune responses and social behaviors, reinforcing the idea that immunity in eusocial species is not solely based on the individual but can extend to colony-level defense strategies (Evans and Spivak 2010).

6.8 Immune Responses to Pesticide Exposure

Pesticide exposure represents one of the most pressing environmental stressors affecting honey bee health. Neonicotinoids, a class of systemic pesticides, are known to affect the central nervous system of insects but also have immune-modulating effects. Sublethal doses of neonicotinoids like imidacloprid have been shown to impair honey bees' immune responses by reducing the expression of critical immune genes such as defensin, a potent antimicrobial peptide, and abaecin, which is vital for defense against Gram-negative bacteria (Di Prisco et al. 2013; Brandt et al. 2017). Furthermore, neonicotinoids may alter hemocyte function, decreasing their ability to effectively phagocytose pathogens. This is a particularly concerning issue, as weakened immune defenses increase the susceptibility of honey bees to viral infections, particularly those transmitted by the *Varroa destructor* mite. Beyond the direct effects on the immune system, neonicotinoids also exacerbate the impact of other environmental stressors such as poor nutrition and climate change, leading to a compounding effect on bee health. As honey bees are exposed to pesticides through contaminated foraging areas, their immune systems become increasingly compromised, reducing their ability to cope with viral infections or pathogens (Desneux et al. 2007). The immunosuppressive effects of pesticides are compounded by the reduced availability of diverse and nutritious forage due to the widespread use of monoculture crops. This dual stressor scenario, in which bees are immunocompromised by pesticides and nutritionally deprived, underscores the need for more sustainable agricultural practices and greater regulation of pesticide use.

6.9 Conservation Strategies and Global Initiatives

In response to the ongoing decline in honey bee populations, various conservation strategies have been proposed and implemented. These strategies aim to protect bee health through habitat restoration, pollinator-friendly farming practices, and the reduction of pesticide use. For example, the planting of wildflowers and other nectar-rich plants has been shown to improve both the nutrition and immunity of honey bee colonies, providing bees with diverse sources of pollen and nectar that are rich in essential amino acids, lipids, and micronutrients (Alaux et al. 2010). Furthermore, integrated pest management (IPM) strategies that minimize the use of harmful chemicals while promoting biological control of pests like *Varroa destructor* can help reduce both chemical exposure and pathogen load in bee colonies. Public awareness and education also play a crucial role in bee conservation. Increasing public awareness about the importance of bees in pollination, food production, and biodiversity can foster greater community involvement in local conservation efforts. Governments and non-governmental organizations (NGOs) have increasingly recognized the need to protect pollinators, implementing measures such as bee habitat corridors, restrictions on pesticide use, and the promotion of organic farming. International initiatives like the International Pollinator Initiative and national-level programs have focused on creating policies that safeguard bee populations, though much work remains to ensure the sustainability of these efforts.

6.10 Future Directions in Bee Immunology Research

Future research in honey bee immunology should aim to bridge gaps in our understanding of how environmental stressors, genetic factors, and immune responses interact at the molecular, cellular, and colony levels. A significant area of focus should be the genetic basis of immune variation among honey bee populations. Understanding how certain colonies or subspecies exhibit enhanced resistance to specific pathogens or environmental stressors could pave the way for breeding programs that select for disease-resistant traits. Additionally, continued research into the molecular mechanisms of the RNAi pathway could offer promising avenues for the development of targeted antiviral treatments for honey bees. Researchers are also exploring the potential of microbiomes in regulating bee immunity. The diverse community of bacteria, fungi, and viruses that inhabit the bee's gut and other body parts may play a crucial role in shaping immune responses. Understanding how these microbial communities interact with the bee's immune system could lead to new strategies for enhancing immune function through microbial management or probiotic interventions (Raymann et al. 2017). Finally, integrating immunological research with ecological monitoring will be vital for assessing the health of bee populations in the field. As climate change continues to alter ecosystems and agricultural practices, it will be essential to monitor the effects of these shifts on bee immunity and colony survival. Long-term data on disease prevalence, pesticide exposure, and environmental factors will help inform policies aimed at mitigating the impact of these stressors on bee health.

6.11 Conclusion

The immune system of honey bees, a cornerstone of their survival, operates through a combination of cellular, humoral, and social immunity mechanisms that provide effective defense against a wide range of pathogens. Despite the complexity of these defenses, honey bees face significant challenges from environmental stressors such as pesticides, climate change, poor nutrition, and the pervasive threat of viral infections, particularly those exacerbated by the Varroa destructor mite. These stressors not only compromise the physiological immune responses of the bees but also interfere with their social immunity, leading to weakened colony resilience and, in some cases, catastrophic colony losses. The intricate interaction between environmental factors and immune function underscores the urgent need for more sustainable agricultural practices and policies that protect pollinators. Honey bee immunity is shaped by both individual immune responses and collective behaviors, such as hygienic behavior, which offers a unique dimension to understanding immunity in eusocial organisms. Furthermore, comparative studies with model organisms like Drosophila melanogaster and Anopheles gambiae provide insights into the evolution and diversification of insect immunity, highlighting both conserved and species-specific immune mechanisms. The global impact of the COVID-19 pandemic, although primarily human-centric, also provided an unexpected opportunity to further explore viral ecology and the implications for bee health, opening new avenues for cross-disciplinary research and conservation efforts. As we look to the future, advancing our understanding of honey bee immunology—particularly the roles of RNA interference, immune gene regulation, and microbial interactions—will be critical in developing novel interventions for enhancing bee health. Conservation strategies, such as habitat restoration, the reduction of pesticide use, and the promotion of bee-friendly farming practices, are essential in reversing the decline in honey bee populations and ensuring their continued role as essential pollinators. By integrating molecular immunology, ecological monitoring, and public engagement, we can create a robust framework for addressing the multifaceted challenges facing honey bee health.

References

Agaisse H, Perrimon N (2004) The Jak-STAT signaling pathway in Drosophila: a paradigm for cytokine signaling. Trends Genet 20(1):10–16

Alaux C et al (2010) Diet quality during early life influences honey bee health. Environ Microbiol 12(8):2134–2143

Blandin S et al (2004) The mosquito immune response to Plasmodium parasites. Insect Biochem Mol Biol 34(7):677–685

Brandt AM et al (2017) Neonicotinoid insecticides and honey bee health: a review of the literature. J Insect Sci 17(1):1–9

Brutscher LM, Daughenbaugh KF, Flenniken ML (2015) Antiviral defense mechanisms in honey bees. Curr Opin Insect Sci 10:71–82

Brutscher LM et al (2017) The role of RNA interference in honey bee antiviral immunity. J Insect Physiol 97:49–57

Casteels-Josson K et al (1994) Apidaecin: a potent antimicrobial peptide from honey bee hemolymph. J Biol Chem 269(22):15300–15306

Cremer S et al (2007) Social immunity. Curr Biol 17(16):R693–R702

Desneux N et al (2007) The sublethal effects of pesticides on beneficial insects: implications for integrated pest management. Environ Toxicol Chem 26(7):1727–1733

Di Prisco G et al (2013) Imidacloprid affects immune responses of honey bees. PLoS One 8(11):e77112

Dicks LV et al (2021) The effects of COVID-19 lockdowns on pollinator research and conservation. Conserv Biol 35(6):1685–1696

Evans JD, Spivak M (2010) Socialized immunity in honeybees (Apis mellifera): evidence for a vital, colony-level defense mechanism. Behav Ecol Sociobiol 64(9):1311–1319

Evans JD et al (2006) Immune pathways in honey bee (Apis mellifera) and other insects: an overview of the innate immune system and molecular immunology. Biochim Biophys Acta (BBA) – Mol Basis Dis 1762(11):1017–1027

Genersch E (2010) Honey bee pathology: current threats to honey bee health. Appl Microbiol Biotechnol 87(2):87–97

Goulson D et al (2015) Bee declines driven by combined stress from parasites, pesticides, and lack of flowers. Science 347(6229):1255957

Hunter WB et al (2010) RNA interference as a mechanism to control honey bee viruses. J Invertebr Pathol 104(3):120–124

Lavine MD, Strand MR (2022) Insect hemocytes and their role in immunity. Insect Biochem Mol Biol 32(10):1295–309. https://doi.org/10.1016/s0965-1748(02)00092-9

Le Conte Y, Navajas M (2008) Climate change: impact on honey bee populations and diseases. Rev Sci Tech 27(2):485–495

Lemaitre B, Hoffmann J (2007) The host defense of Drosophila melanogaster. Annu Rev Immunol 25:697–743
Lemaitre B et al (1996) The Drosophila toll and the innate immune response. Nature 383(6602):102–105
Maori E et al (2009) Honey bee antiviral immunity. Insect Mol Biol 18(5):586–596
Myllymäki H et al (2014) Immune signaling in Drosophila. J Innate Immun 6(4):287–296
Raymann K et al (2017) The role of the microbiome in honey bee immunity and health. PLoS Pathog 13(7):e1006171
Rosenkranz P et al (2010) The impact of Varroa mites on honey bee health: a global perspective. Apidologie 41(3):227–255
Rzymski P et al (2021) Impact of the COVID-19 pandemic on bee research and honey bee health: a cross-disciplinary perspective. J Apic Res 60(2):151–158
Spivak M, Gilliam M (1998) Hygienic behavior of honey bees and its application for control of brood diseases and Varroa mites. Apidologie 29(2):191–210
Spivak M et al (2003) Detection and removal of diseased brood by honey bees: a model for colony-level immune defense mechanisms. Behav Ecol Sociobiol 54(5):453–462
Vallet-Gely I et al (2008) The immune system of insects: a model for studying antimicrobial peptide production and host defense. Curr Opin Microbiol 11(3):165–170

MIX
Papier aus verantwortungsvollen Quellen
Paper from responsible sources
FSC® C105338
FSC
www.fsc.org

If you have any concerns about our products,
you can contact us on
ProductSafety@springernature.com

In case Publisher is established outside the EU,
the EU authorized representative is:
Springer Nature Customer Service Center GmbH
Europaplatz 3, 69115 Heidelberg, Germany

Printed by Libri Plureos GmbH
in Hamburg, Germany